Procedural Justice in the United Nations Framework Convention on Climate Change

Luke Tomlinson

Procedural Justice in the United Nations Framework Convention on Climate Change

Negotiating Fairness

Luke Tomlinson
London, UK

ISBN 978-3-319-38616-4 ISBN 978-3-319-17184-5 (eBook)
DOI 10.1007/978-3-319-17184-5

Springer Cham Heidelberg New York Dordrecht London

Softcover reprint of the hardcover 1st edition 2015

Printed on acid-free paper

Springer International Publishing AG Switzerland is part of Springer Science+Business Media (www.springer.com)

Preface

Multilateral efforts at the global level are yet to produce meaningful action on climate change. In light of this inaction, many have questioned whether the UNFCCC is an appropriate forum for coordinating action, and many alternative arrangements have arisen to fill the regulatory void. Part of this criticism suggests that there is a perceived lack of fairness in the UNFCCC. Whilst academic discussion has traditionally focused on the issue of distributive fairness in this context, very little has been said about procedural fairness. To this end, this book considers what is needed for fairness in the decisions of the UNFCCC. It analyses several principles of procedural fairness in order to develop practical policy measures for fair decision-making in the UNFCCC. This includes measures that determine who should have a right to participate in its decisions, how these decisions should take place and what level of equality should exist between these actors. In doing so, it proposes that procedural fairness is a fundamental feature of a multilateral response to address climate change. By showing that procedural fairness is most likely to be achieved through the inclusive process of the UNFCCC, it also shows that global efforts to address climate change should continue in this forum.

London, UK Luke Tomlinson

Acknowledgements

This book is based on a thesis that was submitted for the Degree of Doctor of Philosophy in Politics, the Department of Politics and International Relations, the University of Oxford in 2014. This was part of the project 'Equity and Climate Change', funded by the Economic and Social Research Council.

I owe a great deal of thanks to my supervisor, Simon Caney, for his guidance and supervision in the earlier stages of my research. I am also very grateful for the feedback provided on earlier versions of my thesis by Jeremy Waldron, David Miller, Elizabeth Fraser and Michael Mason.

I am heavily indebted to several people who generously dedicated their time and thought to reading and discussing my research. I owe special thanks in particular to Clare Heyward and Dominic Roser.

All opinions expressed in this book are my own.

Contents

List of Abbreviations

APP	Asia-Pacific Partnership
AOSIS	Alliance of Small Island States
CDM	Clean Development Mechanism
COP	Conference of the Parties
COP15	UNFCCC Conference of the Parties Copenhagen 2009
COP16	UNFCCC Conference of the Parties Cancun 2010
COP17	UNFCCC Conference of the Parties Durban 2011
COP18	UNFCCC Conference of the Parties Doha 2012
COP19	UNFCCC Conference of the Parties Warsaw 2013
COP20	UNFCCC Conference of the Parties Lima 2014
COP21	UNFCCC Conference of the Parties Paris 2015
G-77	Group of 77 and China
G8	Group of Eight
G20	Group of Twenty
GCCA	Global Climate Change Alliance
IEA	International Energy Agency
IMF	International Monetary Fund
IPCC	Intergovernmental Panel on Climate Change
MEF	Major Economies Forum on Energy and Climate
NGO	Non-governmental Organisation
NSA	Non-state Actor
OECD	Organisation for Economic Cooperation and Development
REDD	Reducing Emissions from Deforestation and Forest Degradation
ROP	Rules of Procedure
UNEP	UN Environment Programme
UNFCCC	UN Framework Convention on Climate Change
WTO	The World Trade Organization

Chapter 1
Introduction

1.1 Introduction

In December 2015, representatives of governments from around the world will meet in Paris to discuss the terms of a legally binding global agreement for action on climate change. This meeting represents the 21st Session of the Conference of Parties (COP21) to the United Nations Framework Convention on Climate Change (UNFCCC), which is the current global arrangement for collective state action on climate change. The ultimate aim of this arrangement is to prevent human interference with the climate from bringing about dangerous levels of climate change. With this end in mind, the purpose of COP21 is to get states to commit to some sort of legally binding agreement on climate change that will come into force in 2020.

Given that climate change is now largely undisputed and a central priority of the global political agenda, COP21 has become the most significant and highly anticipated conference on climate change to date, drawing public attention and inspiring political discussion around the world. As questions over the science of climate change have faded, the urgency of the action needed has become startlingly clear. The result is that COP21 in Paris is now seen as a 'make or break' opportunity for world leaders to act. Provided a new agreement arises in Paris, there will be continued discussions over the final details of this agreement in the coming years. This is just the start of the process towards a comprehensive agreement.

But looking back at previous COP meetings throughout the history of the UNFCCC leaves little room for optimism over the chances of a meaningful agreement arising in Paris. Since its creation over two decades ago, the UNFCCC has brought about little in the way of meaningful action on climate change. Instead, there is an alarming disjuncture between what's collectively required to avoid dangerous climate change and what action has actually been taken so far by individual states, leading to much political, academic and public debate, not only

L. Tomlinson, *Procedural Justice in the United Nations Framework Convention on Climate Change*, DOI 10.1007/978-3-319-17184-5_1

about how to instigate action within the UNFCCC, but also whether it is a suitable forum for addressing climate change at all.

Some authors attribute the lack of action in the UNFCCC to a perceived lack of fairness among its participants, which in turn has created a political stalemate as parties argue over intractable positions. In response to this challenge, many philosophers and political theorists have considered what's fair in relation to climate change. Traditional discussions of fairness in the UNFCCC focus on distributive aspects, which relate to the fair distribution of costs and benefits. Climate change is a both a global and an intergenerational problem for which there is much at stake. Many of those who will experience the very worst effects of climate change will have had little responsibility for bringing it about. There will be many winners and losers in the years to come. As a result, there has always been a strong divide in the COP negotiations between poorer nations, who demand more action from those that have historically caused climate change, and the rich, who expect greater action from rapidly developing economies.

Little, however, has been said about procedural fairness in relation to the UNFCCC, which concerns the way that decisions are made. If distributive fairness relates to the costs and benefits that arise from the actions of the UNFCCC, procedural fairness concerns how decisions about these actions are made. In fact, very little has been said about the procedural rules of the UNFCCC at all. Many of its procedural rules have not yet been formally adopted, operating instead on an ad hoc basis. In particular, the UNFCCC has not yet adopted any rules over voting, with the result that decisions are made by consensus, which allows a single party to obstruct action even if there is agreement among the rest of the group. This has lead to both stalemate and outcomes that consistently reflect the 'lowest common denominator' where ambition and leadership is desperately needed. Further, procedural fairness is not just important in its own right, but also because it can provide a way of reaching agreement amongst those who disagree over distributive issues. Given the stalemate that exists in the UNFCCC and the urgent need for action on climate change, a review and analysis of the UNFCCC's procedural rules is long overdue.

To this end, this book identifies which rules should govern the decisions of the UNFCCC. In specific, it develops a set of rules so that the decisions of the UNFCCC are made *in a fair way.* As such, this book isn't about *what* should be decided in the UNFCCC, but rather *how* its decisions should be made. I argue that if a decision is made in accordance with these rules, then an agreement will gain long-term support and endorsement. In doing this, I also show that, on account of its universal membership, the UNFCCC is the only appropriate forum for international action on climate change. This book therefore makes a twofold contribution to the debate preceding the UNFCCC COP21 meeting in Paris 2015 and its aftermath in the years to come; first, by determining several practical policy measures for instigating action on climate change and second, by arguing that states should continue to support international action on climate change within the UNFCCC. It does this by linking analytical political philosophy with applied public policy.

Given the growing academic and political debate taking place on the future of the UNFCCC, and the need for policy guidance at a practical level, the book provides an important contribution to an otherwise neglected issue area that will be of interest to both academics and practitioners working in the field, including state delegations, NGOs, and international organisations. What's more, many of the arguments in this book will apply to other multilateral agreements to address climate change. Its recommendations will not only guide the UNFCCC in the immediate years to come, but also action on climate change in many other forums.

The book ultimately develops several practical policy measures for the design of the UNFCCC. It does so in four steps. First, it considers the various principles of distributive fairness that have been advocated in the UNFCCC and shows that, not only is there disagreement over what is fair in this context, but also that there is *reasonable* disagreement over how the costs and benefits of climate change should be fairly distributed. This is important because it means that deliberation and discussion is unlikely to bring about an agreed outcome. Second, having shown this, the book then argues that procedural fairness is an important way of reaching a mutually acceptable outcome when there is reasonable disagreement over how to distribute costs and benefits. As a result, the current use of consensus-based decision-making is partly responsible for the political inertia in the UNFCCC and there is a consequent need to revise its procedural rules.

Third, the book develops several principles that should govern the decision-making processes of the UNFCCC, including principles governing who should participate, the terms on which decisions should be made, what voting method should be used, and how actors should bargain. It does this by analysing principles of procedural fairness and considering how they can be applied in the context of climate change. Finally, it argues that the UNFCCC is the only appropriate forum for addressing climate change at the global level. It does this by arguing that procedural justice is a fundamental part of any effective climate change agreement. Drawing on the earlier arguments, the book suggests that procedural justice requires that a fair climate agreement have universal representation. Given that the UNFCCC is the only forum that provides universal membership, it is the only appropriate forum for effectively addressing climate change. As such, the book serves as both an academic study of procedural justice and climate change as well as a guide for policy-making for international cooperation on climate change.

1.2 Climate Change and the UNFCCC

The fact that the earth's climate is undergoing fundamental changes due to human activity is now undisputed (UNFCCC 1992; IPCC 2007, 2012; Stern 2007; Garnaut 2009). The Assessment Reports of the Intergovernmental Panel on Climate Change (IPCC) are the most comprehensive studies to date on climate change and the impacts that it has on human interests.[1] The most recent of these, the IPCC Fifth Assessment Report published in 2013, represents the most comprehensive study of the climate ever undertaken (IPCC 2013). This report states that climate change is occurring, that this is very likely due to human activity, and that unabated action will result in further climate change (IPCC 2014 Synthesis Report). The potential implications of climate change include severe and irreversible changes to the climate system, which are expected to have extreme consequences for fundamental human interests on a global scale.[2] This includes sea level rise and an increase in the incidence of extreme weather events such droughts. It is widely thought that this will threaten basic human rights to food, water, health and shelter and could represent an existential risk to some countries (Caney 2009; OHCHR 2009; OHCHR and UNEP 2012).[3]

The IPCC also states that climate change mitigation is a global commons problem, for which collective action to reduce emissions of greenhouse gases will provide greater aggregate gains than continued unrestricted emissions (Toth et al. 2001, p. 653; IPCC WG3 TS.4.4). As such, the potential benefits of avoiding severe climate change are expected to outweigh the anticipated costs of achieving this objective (IPCC 2007; Stern 2007; Garnaut 2009).[4] This has been reiterated by the Global Commission on the Economy and Climate, a group of experts, commissioned to analyse the economic implications of addressing climate change (the Global Commission on the Economy and Climate 2014).

Given that climate change is a global problem and that no single actor is responsible for a significant proportion of total emissions, achieving climate mitigation is often seen as requiring an international, if not fully global response (IPCC 2001, 2007). The United Nations Framework Convention for Climate Change (UNFCCC) is the existing international agreement for international cooperation on climate change and the Conference of the Parties to the UNFCCC (COP) is the official negotiating forum for collective decision-making in the Convention (UNFCCC 1992).[5] This is an international agreement among nation states to cooperate on climate change. The ultimate objective of the Convention, which and has been signed and ratified by 196 states, is to 'stabilise greenhouse gas concentrations in the atmosphere at a level that prevents dangerous interference with the climate system' (UNFCCC 1992, Article 2).

---

[1]The remainder of this book draws from the Assessment Reports of the IPCC, in particular: Banuri et al. 1995; Bashmakov et al. 2001; Toth et al. 2001; Gupta et al. 2007; Halsnæs et al. 2007; Stavins et al. 2014.

[2]For further discussion, see: IPCC 2007; Caney 2009; OECD 2012.

[3]Small island nations are particularly vulnerable to sea level rise. See: Yamano et al. 2007.

[4]There is not complete consensus on this matter. Some argue that it would be better to pursue other policy options aside from mitigation. See, for example: Schelling 1997; Lomborg 2001; Nordhaus 2009.

[5]For discussion: Bodansky 1993, 2001; Yamin and Depledge 2004; Depledge 2005.

The legal instrument of the UNFCCC is the Kyoto Protocol (1997), which puts legally binding commitments on states to reduce their greenhouse gases (UNFCCC 1997). The Marrakech Accords (2001), the Bali Action Plan (2007), and the Durban Platform (2011) are subsequent agreements that have been adopted to continue action through the UNFCCC. Whilst the recent COP15 negotiations in Copenhagen highlighted the limits of the UNFCCC process, the outcome of COP16 in Cancun and COP17 in Durban renewed optimism in its ability to deliver collective action on climate change.[6] In particular, COP17 established a second commitment period under the Kyoto Protocol, as well as the Ad-hoc Working Group on the Durban Platform: an agreement to negotiate an agreed outcome with legal force by 2015, which will become operational in 2020. The recent COP18 in Doha (UNFCCC 2012) committed to build on the framework put in place at Durban and this process was reaffirmed at COP19 and COP20 in Warsaw 2013 and Lima in 2014 respectively (UNFCCC 2013; UNFCCC 2014).

Although there is some dispute over what dangerous climate change exactly entails, avoiding dangerous anthropogenic interference with the atmosphere is now broadly seen as limiting global temperature increases to within 2 °C of those before the industrial revolution. The IPCC states that in order to keep a 50 % chance of meeting this target, it is necessary to limit atmospheric concentrations of green house gases to between 480 ppm and 530 ppm, which in turn requires drastic reductions in the overall levels of global green house gas emissions. But little action is taking place to mitigate the activities that cause climate change and the international community has struggled to come up with a collective response to this problem. On current trends, temperature increases could exceed 4 °C by the end of this century, which would lead to extreme and irreversible impacts (Global Commission on the Economy and Climate 2014). Some think that the lack of action achieved by the UNFCCC requires a major reassessment is needed of the current focus to implement action through the UNFCCC. It would be more worthwhile pursuing international action in other international forums and to focus attention elsewhere.

There are many other multilateral arrangements that coordinate cooperative action on climate change where it might be much easier to stimulate action.[7] These are agreements amongst limited numbers of states to address climate change, including traditional international institutions that are now incorporating climate change into their mandates, such as: the Group of Eight Industrialised Countries (G8), the Group of Twenty (G20) and the UN Security Council. Given that the G8

[6]For criticism of COP15, see: Dubash 2009, p. 8; IISD 2010; Winkler and Beaumont 2010, p. 640. For discussion of COP16, see: King et al. 2011. For commentary on the Durban Platform, see: Fu-Bertaux and Ochs 2012. For the outcomes of COP17, see: UNFCCC 2011.

[7]For more on climate change initiatives outside of the UNFCCC, see: Jagers and Stripple 2003; Pattberg and Stripple 2008; Biermann 2010; Bulkeley and Newell 2010.

and G20, as well as agreement set up to specifically address climate change such as the Major Economies Forum on Climate Change and Energy (MEF).[8]

There are also many arrangements between state and non-state actors at the international and national level.[9] Examples include the Netherlands Voluntary Agreement on Energy Efficiency and the Australia Greenhouse Challenge Plus Program (Gupta et al. 2007, p. 761). National laws and policies are also critical areas of climate policy (Levi and Michonski 2010). For this reason, some authors argue that power over collective action for climate change is increasingly located beyond the intergovernmental system (Kingsbury et al. 2005; Pattberg and Stripple 2008; Biermann et al. 2009; Biermann 2010; Corbera and Schroeder 2011). Others argue that climate change politics is decentralised, or 'fragmented', reflecting the multiplicity of actors and power relations that exist beyond the traditional interstate system (Biermann et al. 2010). As such, the failure of centralised approaches to action on climate change, and the increasing prevalence of alternative forms of cooperation, has lead some to suggest that action might be better pursued in forums outside of the UNFCCC (Prins and Rayner 2007a, 2007b; Grasso and Timmons Roberts 2013).

These different international arrangements aren't mutually exclusive, and many work alongside one another. However, focusing international efforts to address climate change in one arena does limit the resources that can be put into achieving outcomes in other areas, so there are tensions between these different forums for cooperation. For one thing, the UNFCCC can be perceived as the overall institution that should deliver action on climate change, so waiting for a top down agreement to arise may prohibit action in other areas as states anticipate action to come about. The costs of the annual COPs aren't insignificant either. Given what's at stake, the lack of action active by the UNFCCC thus far, and the emerging diversification of alternative arrangements for international cooperation, it's worth considering whether the UNFCCC is still the most appropriate forum for addressing climate change.

1.3 Guiding Principles for International Cooperation on Climate Change

In light of the different institutional arrangements and institutions that exist in relation to climate change, a number of authors have evaluated the UNFCCC and proposed options for its reform.[10] Many of these evaluations and proposals are based

[8]The MEF facilitates dialogue among 17 countries (MEF 2013).

[9]For a description of public-private agreements, see: Gupta et al. 2007, p. 761. For discussion, see: Bulkeley and Newell 2010; Bäckstrand 2008.

[10]This section draws on the discussions from: Höhne et al. 2002, p. 34; Aldy et al. 2003; Bodansky and Chou 2004; Aldy and Stavins 2007, 2010.

on implicit assumptions about the normative desirability of different arrangements and the role that they should play. The IPCC Fourth Assessment Report defines several principles and criteria that can be used to either evaluate existing cooperative arrangements or guide their design (IPCC 2014). Typically, the overall desirability of an institution relates to its performance, or 'effectiveness' in reaching an overall objective (for example, achieving climate stabilisation). But many refer to other normative criteria when making proposals about climate institutions, including justice, legitimacy, and economic efficiency.

The literature on climate change typically divides normative criteria into two categories: substantive criteria, which relate to the outcomes of an institution, and procedural criteria, which relate to the processes that generate these outcomes. These criteria are interlinked, in the sense that they can either complement and conflict with one another in different situations. For example, an institution that achieves economic efficiency may not yield the best environmental outcome (Philibert and Pershing 2001). On other occasions they are mutually supportive; an institution that is neither equitable, nor politically feasible, is unlikely to achieve its goals (Rajamani 2000).

Substantive criteria can also relate to the procedural design of an institution, just as procedural criteria can be matters of substantive concern. For instance, it might be desirable to design an institution that achieves a substantive end, such as economic efficiency. This involves ensuring that the outcomes of the agreement are those that minimise the economic cost of the agreement. But it also involves designing the procedural aspects of the institution so that these minimise the economic cost of the agreement as well. This might involve designing procedures that minimise transaction costs, or that do not place high information costs on participating actors. In this instance, a substantive normative criterion has implications for the procedural design of the institution. Consequently, whilst a distinction can be made between procedural and substantive criteria of institutional design, this does not limit the aspects of institutional design that each type of criterion applies to.

Further, some of the criteria proposed here have *both* procedural *and* substantive elements. For instance, the criterion of legitimacy, which is defined and discussed in the following section, has elements that govern both of these criteria. Separating these elements is common in the literature on institutional design. For example, Fritz Scharpf has labeled these 'input legitimacy' and 'output legitimacy', which respectively relate to procedural and substantive elements (Scharpf 1999). Both Thomas Franck, and Buchanan and Keohane also make a distinction between these two features of legitimacy (Franck 1995; Buchanan and Keohane 2006). The reason for separating such criteria into their substantive and procedural components, even when some of these criteria concern both of these dimensions, is to show that, in certain cases, they matter to both process *and* outcome. This is something that is sometimes overlooked in the literature. I separate these two features to demonstrate that one can focus on the procedural aspect of legitimacy irrespective of the substantive ends that it brings about.

Substantive criteria relate to the outcomes that are brought about by the institution. These criteria are important regardless of the process through which they arise. For instance, one might argue that achieving important ends such as avoiding dangerous climate change is the most pressing concern at the moment and that it doesn't matter how this end is actually achieved, so long as this goal is reached. The literature on multilateral climate change institutions often refers to five substantive criteria for institutional design: effectiveness; justice, or equity; efficiency, or cost-effectiveness; legitimacy; and political feasibility.[11] These criteria are valuable in themselves, but they are also interdependent and interlinked.

Effectiveness relates to the extent to which an institution meets its intended objective (Höhne et al. 2002, p. 33). In the case of climate change, this is typically defined in terms of meeting an emissions target or achieving certain adaptation goals. For example, the primary objective of the UNFCCC is to stabilise greenhouse gas concentrations in the atmosphere at a level that prevents dangerous interference with the climate system (UNFCCC 1992, Article 2). This is often proposed as the primary objective of any international agreement, and many authors advocate secondary principles that are instrumental to achieving this end. For instance, some argue that a high level of participation is a fundamental criterion of institutional design for climate change policy (OHCHR 2009, p. 23; Hare et al. 2010; Bosetti and Frankel 2012). Others claim that compliance and enforcement are essential elements of a multilateral climate change agreement (Barrett 2003; Barrett and Stavins 2003; Victor 2006). However, to a large extent, participation and enforceability are only desirable insofar as they are instrumental towards achieving emissions reductions or some other end.

In addition to effectiveness, there are three further substantive principles that are often given as guiding principles for climate policy: efficiency, legitimacy and distributive equity. Efficiency dictates that the economic costs of addressing climate change are minimised (Gupta et al. 2007, p. 750). Broadly speaking, a legitimate institution is one that has both the right to govern as well as a level of support amongst those on whom it imposes power (Franck 1995; Dingwerth 2005; Bodansky 2007). Distributive equity relates to both justice and fairness and concerns how the relative benefits and costs of climate change should be distributed amongst states. There are other elements of equity that are important in relation to climate change, including intergenerational equity, or intranational equity. However these elements are beyond the scope of this book. Further, whilst equity, justice and fairness sometimes have different meanings in different contexts, this isn't such a concern for the content of this book. Here, following the convention of much of the literature on this subject, I use these terms interchangeably.[12]

But the design of a multilateral climate change agreement is not simply a matter of promoting certain substantive outcomes: procedural values also have a role to

[11]Gupta *et al.* refer to these as 'desirable' criteria (Gupta et al. 2007, p. 750). These are also referenced in other IPCC Assessment Reports (see: Bashmakov et al. 2001, p. 407). Other authors who refer to these criteria include: Aldy et al. 2003, p. 374; Aldy and Stavins 2010, p. 2–3; Winkler and Beaumont 2010, p. 642.

[12]Here, I follow: Toth et al. 2001, p. 668, footnote 40.

play here. Procedural values are those that relate to how an outcome is reached, regardless of what that outcome actually is. When thinking about climate change, this concerns the design of decision-making processes that determine outcomes, which are the institutional procedures for making collective choices (Krasner 1982, p. 186).

There are two procedural normative criteria that are of particular concern for institutional design: procedural efficiency and procedural justice. On the one hand, procedural efficiency relates to the ability to actually make a decision. That is, it concerns how issues such as how easy it is for a group to agree on something. There are various proposals for facilitating decision-making in multilateral climate change institutions, including:

1. Simplifying decisions and limiting separate discussions.[13]
2. Ensuring that decisions are compatible with prior convention.[14]
3. Splitting up problems into smaller negotiation packages.[15]
4. Designing negotiations to focus on the problem at hand.[16]

Procedural justice, on the other hand, concerns whether the means by which an outcome is reached is fair regardless of what the outcome actually is (Banuri et al. 1995, p. 83–5, 117; Rayner and Thompson 1998, p. 319; Albin 2001; Grasso 2007; 2010, p. 4). It relates to who participates in a decision-making process, as well as the fairness of that process. The basis of procedural justice is grounded in different ways according to different theories of justice. Whilst some argue that procedural justice is based on a fundamental duty of equal respect for the opinions of others,[17] others claim that procedural justice is important because it enables affected parties to maintain their dignity (Schlosberg 1999, p. 12–13, 90; Paavola 2005, p. 313–4), or that it carries important instrumental value towards meeting other ends (Toth 1999, p. 2).

Whilst many authors acknowledge the importance of procedural fairness, this issue is often overlooked in the literature on climate change, and formal mechanisms to facilitate procedural justice are missing from most policy proposals.[18] This is strange given the importance that many place on the value of procedural justice in political institutions generally, as well as in specific relation to climate change. The climate negotiations at COP15 highlighted that the COP is seen as an illegitimate venue for negotiations due to the exclusive nature of its decisions. At these negotiations, Venezuela, Cuba, Nicaragua and Bolivia all renounced the Copenhagen

[13]Torvanger and Ringius 2002, p. 224; Höhne et al. 2002, p. 34; Ghosh 2010, p. 3.

[14]Depledge and Yamin 2009.

[15]Biermann et al. 2011.

[16]Gupta et al. 2007; Harstad 2009. Arunabha Ghosh also discusses some features that may improve multilateral negotiation dynamics (Ghosh 2010, p. 4).

[17]For example: Waldron 1999.

[18]Exceptions include: Adger et al. 2006; Grasso 2010.

Agreement on procedural grounds.[19] Others have questioned the legitimacy of the G8 and MEF on procedural grounds, arguing that they exclude key actors and are insufficiently transparent (Karlsson-Vinkhuyzen and McGee 2013, p. 67).

To be sure, some authors argue that there is no intrinsic merit to procedural design and that decision-making processes should be designed with the sole intention of promoting certain desirable outcomes. For instance, Richard Arneson argues democracy should be regarded as 'a tool or instrument that is to be valued not for its own sake but entirely for what results from having it' (Arneson 2004). In specific relation to climate change, one might hold that the overall goal of international cooperation should be to avoid dangerous climate change, and that other values should only be taken into account to the extent that they promote this end. According to these sorts of arguments, decision-making processes should be designed to achieve desired outcomes, rather than to promote values that are independent of these ends.

But in response to this objection, there are at least four reasons for considering procedural values in the context of climate change. First, as mentioned above, many authors do in fact argue that there is something intrinsically just about the process by which outcomes are reached.[20] This view is supported by empirical studies of human behaviour, as well as the claims that arise in the negotiations of the UNFCCC (IISD 2010; Grasso 2010, p. 99; Winkler and Beaumont 2010, p. 640). It seems reasonable to assume that there may be some cases in which the process is valued independently of the outcome achieved.

Second, the absence of procedural justice in the UNFCCC has contributed to its political deadlock. Despite several unsuccessful attempts to adopt rules for voting and broader decision-making, the Parties of the COP have so far failed to agree on its procedural rules, meaning that the COP continues to rely on the draft rules of procedure, which do not specify a voting procedure.[21] Consequently, decisions are made by consensus, leading to 'lowest common denominator' outcomes, and blocking tactics within negotiations (Prins and Rayner 2007b).

Third, as I show in Chap. 2, climate change is an issue that is characterised by reasonable disagreement over substantive values. This means that no single answer is likely to gain the support of actors, even if they are arguing about the common good, and in good faith. I show that procedural justice is instrumentally important because it allows us to reach a mutually acceptable outcome when there is reasonable disagreement about substantive ends. In Chap. 8, I go on to argue that procedural justice is still important even if people give up on a comprehensive agreement for climate change.

Fourth, climate change is characterised by extreme uncertainty, and this makes it difficult for actors to reach agreement on the substantive outcomes of climate

[19]See: Dubash 2009, p. 8; Bäckstrand 2010, p. 1.

[20]Those who argue that there is an intrinsic value to democracy include: Beitz 1989; Cohen 1997; Waldron 1999.

[21]For the UNFCCC Rules of Procedure on voting, see: UNFCCC 1996, Rule 42.

institutions. Some authors argue that, when outcomes are uncertain, actors put an emphasis on the quality of the procedure, over substantive outcomes (Toth 1999, p. 2; Foti et al. 2008). For this reason, procedural justice is an important precondition to creating and operating climate institutions.

Regardless of what one thinks about the intrinsic value of procedural justice there are good reasons for thinking about the procedural fairness of the UNFCCC and climate policy more generally. In fact, one of the central arguments of this book is that procedural justice is a fundamental element of effective climate change institutions. This book develops this idea further and uses it to develop a set of principles for fair decision-making in the UNFCCC.

1.4 Literature Review

Before explaining how this will be done, it's worth reflecting on the existing literature on procedural justice and the UNFCCC. Whilst there is a great deal of research on the role of justice and morality in climate change, many of these studies almost completely overlook procedural justice. Theorists and philosophers have examined issues of fairness and justice extensively since climate change became a matter of political concern.[22] Debate has largely focused on: which principles should guide the distribution of the benefits and burdens of climate change; the fair distribution of emission rights; what rights and duties people have regarding climate change; and, to a limited extent, fair adaptation to climate change. Whilst many authors acknowledge that procedural justice is an important issue in this area, few political theorists have taken up this question.

There has, however, been a recent 'procedural turn' in the literature on multilateral governance, as authors have turned their attention to issues of inclusiveness and transparency (Bäckstrand 2006, p. 467; Bäckstrand 2010; Dryzek and Stevenson 2011, 2012b; Stevenson 2011). There is growing consensus in both the academic and policy literature that principles of 'good' governance should apply to decisions in international organisations. Core elements of good governance include: transparency, participation, accountability, and the review and refinement of policy choices over time. Other studies specifically consider the fairness of international negotiation processes (Albin 2003, p. 13; Chasek and Rajamani 2003), often suggesting that the effective representation of all stakeholders and the impartial consideration of all claims are necessary conditions for fair negotiations, or that parties should have more equal starting positions in terms of negotiating capacity.

This procedural turn in literature on multilateral governance is reflected in the gradual accommodation of procedural matters in the treaty texts and constitutions of several multilateral institutions. For example, the Aarhus Convention on Access to Information, Public Participation in Decision-Making and Access to Justice in

[22] Prominent accounts include: Shue 1992; Caney 2006; Miller 2008.

Environmental Matters (1998), which assign rights to information, participation and accountability in environmental decision-making.[23] Authors increasingly recognise that democratic values, such as representation, deliberation and inclusion are becoming common elements of the rhetoric of many other multilateral institutions (Bäckstrand 2010, p. 670).

These values are typically underpinned by a substantive right to live in an environmental adequate for health and well-being. Alternatively, they are advocated as instrumental values for the achievement of other ends. For instance, Karen Bäckstrand argues that the institutional effectiveness of environmental governance mechanisms is tied to procedural values such as representation, participation, accountability and transparency (Bäckstrand 2006, p. 468). As such, the promotion of these democratic values is instrumental towards achieving substantive ends, rather than intrinsically valuable. These studies often consider multilateral institutions more generally, rather than climate change institutions in specific. Whilst climate change institutions are a part of multilateral institutions more generally, there are also many important differences between climate change institutions, and other environmental agreements. Therefore, despite this procedural turn in environmental governance, more attention is needed on the procedural design of the UNFCCC.

Whilst specific references to procedural justice are often absent from the literature on climate change, this is not to say that procedural justice has not featured in the analysis of institutions that share many of the features of those that concern climate change. An example of this is the broad literature on the 'democratic deficit' that exists in global institutions.[24] Following the increased interdependence of the new global order, there has been a revaluation of several notions of legitimacy and democracy in international politics. This reflects the idea that there is now a high level of interdependence between individuals in different countries, and that a democratic deficit has arisen due to the 'transfer of decision-making authority away from nation states towards a variety of unelected or unaccountable international bodies' (Bodansky 2007). As a result, many argue that greater levels of inclusion should take place in the decision-making processes of international institutions,[25] or that there should be greater means through which accountability can be exercised in international institutions.[26] The areas of transnational democracy, cosmopolitan democracy, discursive democracy and stakeholder democracy are all advocated as institutional innovations that reduce the democratic deficit of multilateral institutions in this way (Bäckstrand 2008, p. 79).

But despite this literature, there are several reasons why it's still necessary to consider procedural justice specifically in relation climate change. For one

[23]For the Aarhus Convention, see: UN 1969, 1998.

[24]For example, see: Chimni 2004; Kingsbury et al. 2005; Habegger 2010.

[25]See: Keohane 2003, p. 132; Held 2004, p. 66; Besson 2009, p. 64.

[26]Zürn 2000; Nye 2001; Keohane 2006.

thing, it isn't clear whether such arguments are appropriate in the context of climate change. Climate change is a unique phenomenon, which is separate and distinct from other issues areas. The fact that climate change is global, potentially catastrophic, and highly complex necessitates a new inquiry on the subject of democracy and global politics.[27] For another matter, arguments based on concerns about 'democratic', or 'accountability' deficits are not confined to the fairness of the decision-making processes of multilateral organisations. Rather, these arguments are based on notions of legitimacy, or the need for democratic processes. Whilst recognising the importance of these studies, this book attempts to address procedural design in climate change institutions from the point of view of procedural justice. Furthermore, whilst there is significant debate regarding procedural, or democratic, principles on a global context, this often operates at a high level of abstraction and does not focus on concrete principles of procedural justice.

To be sure, there are some studies that specifically address procedural values and institutional design the context of climate change. The recent collections of studies by Aldy and Stavins, go some way to providing a comprehensive analysis of both procedural, and substantive normative criteria for the institutional design of a successor to the Kyoto Protocol (Aldy and Stavins 2007, 2010). These studies examine the merits of different institutional architectures and make policy proposals for the design of new multilateral agreements. For example, Bard Harstad proposes seven bargaining rules that would facilitate agreement in climate negotiations, including rules regarding minimum levels of participation in bargaining processes, and unanimity requirements for acceptable bargains (Harstad 2010). Other authors have also started to consider the importance of procedural values for the design of climate institutions. For instance, Jing Huang argues that the decision-making processes of a climate regime should be transparent and operate in accordance with established international norms, principles, and laws (Huang 2009). More recently, the Nordic Council of Ministers released a paper that criticised the procedural design of the UNFCCC (Vihma and Kulovesi 2012). The Assessment Reports of the IPCC also provide recommendations for the design of decision-making procedures in the UNFCCC (IPCC 1990; Banuri et al. 1995, p. 57–8).[28]

However, to a large extent, these studies provide an inadequate analysis of procedural justice. The rules proposed by Harstad, the IPCC and the UNFCCC are designed to facilitate agreement among parties, rather than to promote procedural justice. As such, these studies represent proposals for procedural effectiveness, and do not recognise the importance of procedural justice.

There are two important exceptions to the absence of procedural justice in the literature on climate change. Marco Grasso has recently applied principles of

[27]For an argument that climate change is a unique problem, see: Toth et al. 2001, p. 603.

[28]Note that the reason for not using the most recent IPCC reports for citation is that many of the arguments referenced here are covered in earlier IPCC Reports and the authors of more recent reports cite these older publications.

procedural justice to the context of the funding of adaptation to climate change impact (Grasso 2010, p. 38). Grasso develops a theory of procedural justice that is characterised by moral reciprocity and gives a critical review of the current institutional arrangements for international adaptation funding. Huq and Khan also provide a similar analysis, examining the National Adaptation Programmes of Action (NAPA) process in Bangladesh and the role of equity and justice in the decision-making procedures of this policy (Huq and Khan 2006, p. 189). Further to these two exceptions, several authors argue that it is necessary to consider procedural justice in the context of climate change, whilst not exploring the issue in any depth.[29] For example, Adger et al. argue that issues of participation and representation are key concerns regarding adaptation and climate change, yet fail to show why such issues should be taken into account or to demonstrate how they compare to other criteria such as substantive justice.[30] Whilst these studies represent the most thorough analyses of procedural justice in relation to climate change to date, they are limited to specific issue areas of climate change policy and do not address the procedural justice issues within the overall policy architecture of the UNFCCC.

Another area in which procedural issues are addressed in relation of climate change is that of deliberative democracy. While there are differences between different conceptions of deliberative democracy, there is relative agreement that deliberative processes are characterised by a desire to achieve a consensus through open and reasoned argument, free from manipulation and the exercise of power.[31] Deliberative democracy is studied in the context of international environmental politics in general, as well as more specifically in relation to climate change.[32] The recent proliferation of research on this topic is coined the ‘deliberative turn’ in environmental politics, reflecting the increased attention to deliberative qualities in this area.[33]

Whilst deliberative ideals may share some of the features of procedural justice, the two topics remain distinct. Deliberative democracy may be advocated on the basis that it produces desirable, or effective, outcomes rather than for any considerations of fairness. It is possible to promote rules that drive deliberative discussion without necessarily encapsulating any ideas of fair process. Further to this point, as I will show in Chap. 2, climate change is likely to be characterised by disagreement that cannot be resolved through deliberation alone, and it is necessary to consider the process of voting as a non-deliberative procedure for collective decision-making.

[29]For example: Banuri et al. 1995, p. 117; Paavola 2005.

[30]Adger et al. 2003.

[31]See: Cohen 1989, p. 91; Benhabib 1996, p. 68; Gutmann and Thompson 1996.

[32]For studies on deliberative democracy in relation to climate change, see: Dryzek and Niemeyer 2006; Dryzek and Stevenson 2011, 2012a, 2012b; Stevenson 2011.

[33]Bäckstrand et al. 2010.

As such, whilst there is some discussion of procedural justice and climate change in the literature, these studies are either insufficiently normative or provide inadequate focus on procedural justice. But before proceeding, it's worth briefly reflecting on why procedural justice hasn't been discussed more extensively in this context. Since climate change has only recently become a concern in a policy context, it might be that normative theorists have not been able to address procedural justice at this time. Yet the recent nature of the climate problem has not prevented the growth of a vast literature on distributive justice and climate change. Furthermore, the topical nature of the issue has encouraged a great deal of research on the institutional design of international climate change institutions. Given the existing literature on distributive justice and the distribution of greenhouse gas emissions, it could be that the unresolved nature of this debate has prevented progress on procedural justice, as theorists have focused on substantive concerns. Further, some suggest that the lack of attention to procedural justice may be due to the fact that the UNFCCC sufficiently addresses the issue already.[34] Although having said this, given the problems outlined so far, this is clearly not the case.

1.5 Methodology

The purpose of this book is to identify practical policy recommendations that can be used for the design of the UNFCCC and other international climate change institutions. Having outlined the motivation for doing this and having considered the existing literature on the subject, it's time to think about the method that this book will use to achieve this goal. This book identifies normative principles for institutional design using a common methodology in analytical political theory for determining moral judgments, namely: the method of Wide Reflective Equilibrium. Further, developing normative theories of justice typically involves making various assumptions about how actors behave and how the world works in practice. This is often defined in terms of 'ideal' and 'non-ideal' theory. There are already many accounts and discussions of both the method of Wide Reflective Equilibrium and ideal and non-ideal theory that provide a far more detailed and thorough description of this methodology, as well as its strengths and shortcomings, than could ever be provided in the scope of this book (Rawls 1971; Daniels 1996). Still, given that this book is aimed at a wide audience it's worth briefly outlining these topics here for the sake of clarity. To this end, this section provides a brief account of the method of Wide Reflective Equilibrium and outlines some of the assumptions that it later makes.

[34]Klinsky and Dowlatabadi 2009, p. 96.

1.5.1 Wide Reflective Equilibrium

In order to define a Wide Reflective Equilibrium, it's necessary to first say something about what a Reflective Equilibrium entails.[35] A Reflective Equilibrium is a method that can be used to derive moral or non-moral judgements. This takes two forms. A Narrow Reflective Equilibrium involves considering different moral judgements about particular situations in order to identify principles that link those moral judgements to others that are similar. However, in order to find adequate reasons for accepting these moral judgements, it is necessary to use a second type of Reflective Equilibrium, that is: a Wide Reflective Equilibrium. This involves working back and forth among judgements about moral principles and other theoretical considerations in an attempt to find coherence between three sets of beliefs: moral judgements, moral principles and relevant background theories.

This goes as follows: first, a person's moral judgements are considered. Alternative sets of moral principles that fit with the moral judgements are then considered and arguments are advanced to compare the relative merit of alternative principles and competing moral conceptions. These arguments are taken as inferences from the set of relevant background theories. People then make adjustments and revisions between these moral judgements, moral principles and background theories until they reach an equilibrium point. People reach this equilibrium point, or 'reflective equilibrium', when they have an acceptable coherence among these beliefs.

Some argue that this gives undue prominence to moral intuitions. Such moral intuitions may be tainted by historical accident or prejudice and, in this sense, we should be cautious about using these as starting points for thinking about what justice requires. Judgements about justice may be 'local' and specific to certain societies and contexts.[36] Given this, we might think that we shouldn't rely on such an unreliable method that merely reflects a set of determinate moral judgements. But to take this view is misunderstand the point of this method. Rather than simply systematising some determinate set of judgements, the method of Wide Reflective Equilibrium permits extensive revision of these judgements in light of our understanding about the relative merit of different moral principles. Wide Reflective Equilibrium therefore allows for the fact that some principles may be relative, whilst remaining an important undertaking. As such, this book uses this methodology to identify principles of justice that can be used to think about what's required for fairness in the UNFCCC.

A frequently heard complaint about using this sort of method to inform practical policy measures is that, whilst theorising about principles of fairness is all well and good in an academic context, the realities of life makes this work irrelevant for thinking about what's necessary in the real world. Why should policy-makers give

[35]The following account is based on the discussions in: Rawls 1971; Daniels 1996; McDermott 2008.

[36]Daniels 1996, p. 103.

any attention to philosophical arguments about what is or isn't fair in the UNFCCC? Surely it would be a more useful exercise to just find a practical and pragmatic way of getting action on climate change. If states quote principles of fairness in climate negotiations, then it might be better to just accept these as they are, rather than worry about their ethical merit and conceptual clarity. But whilst it's always important to justify one's approach to this sort of exercise, dismissing any consideration of fairness seems ill conceived. It's clear that state delegations do care a lot about fairness in the UNFCCC. Claims about what is or isn't fair have always dominated COP negotiations, and the fact that so much rests on these claims certainly makes analysing the merit of these claims worthwhile.

1.5.2 Ideal and Non-ideal Theory

But in order to think about what fairness requires in the UNFCCC, it's necessary to say something about the assumptions that we'll make along the way. Studies to determine theories of justice typically distinguish between two types of analysis in this respect: 'ideal' and 'non-ideal' theory.[37] Whereas ideal theory defines principles at a fundamental level, making assumptions for the sake of developing a theory of justice, non-ideal theory, on the other hand, takes into account the real world considerations that are excluded from ideal theorising. Whilst ideal theory gives us aspirational goals to aim for, non-ideal theory puts emphasis on what's immediately achievable.

There is an on going debate between those who invoke ideal hypotheticals when deriving principles of justice and those who stress the importance of non-ideal considerations.[38] The concern amongst some is that ideal theories of justice leave us with principles that do not take into account important features of the world, and that the principles prescribe unattainable ends. The counterclaim is that focusing on non-ideal theory leads to a theory of justice that merely reflects what is palatable, or generally acceptable, given people's political motivation for justice. This presents a dilemma for anyone considering normative principles of justice. On the one hand, it is necessary to take into account the existing state of affairs to suggest reforms that are realisable in practice. On the other hand, it is important not to lose sight of the more radical reforms that people should ultimately achieve.

This dilemma is less problematic than it first appears. It is possible to make policy proposals that are achievable whilst maintaining a focus on an ultimate ideal. What is important, however, is that the sort of analysis that is being undertaken is clearly defined. This book recognises the merit of ideal principles whilst realising that such

[37]Buchanan 2004, p. 55; Mason 2004, p. 265; Ypi 2010, p. 538.

[38]For more on this discussion, see: Mason 2004, p. 265; Swift 2008, p. 365; Hamlin and Stemplowska 2012, p. 51.

principles may have to be adapted in order to gain traction in a policy context. For this reason, it's necessary to spell out some of the key assumptions on which the policy prescriptions of this book are made.

First, I assume that states will broadly comply with the principles and commitments of the UNFCCC, provided that it is universally seen as fair. This is a contentious assumption to make, especially in light of the absence of a global sovereign power and the subsequent lack of effective compliance mechanisms for enforcing regulation, which represents a serious challenge to this approach.[39] In response to this challenge, it's important to note that states do participate in, and comply with, international agreements for reasons aside from self-interest. Abram Chayes and Antonia Chayes argue that states comply with the commitments of international agreements because they have been persuaded to do so in the negotiation process, rather than out of a concern for self-interest.[40] Similarly, Thomas Franck argues that fairness is the key determinant of compliance in an international regime (Franck 1995, p. 26).[41] Whatever the primary motive for state action may be, it is clear that climate change negotiations are part of an ongoing cooperative process that take place against the backdrop of many other multilateral arrangements. For this reason, it seems reasonable to assume that states have incentives to join and comply with the provisions of multilateral agreements, even in the absence of a coercive authority.

But there is a second response to this challenge, which is to say that we're developing principles of justice in an ideal sense. Ideal theory involves making certain assumptions about the state of the world. One such assumption is that actors will comply with principles of justice. Whilst there may be problems of non-compliance in a non-ideal context, these issues can be put aside for the moment and later revised against non-ideal scenarios. What matters is to specify what justice requires first, before considering how to deal with non-ideal matters.

Further, this book adopts an 'institutionalist' approach to climate policy, taking institutions as they are and attempting to implement reform on the basis of the current state of the world.[42] Some authors think of this as a 'bottom-up' approach, which attempts institutional reform given the existing state of the world. A non-institutionalist approach, on the other hand, asks what sort of institutions should be created regardless of those that are currently in existence. This is sometimes referred to as a 'top-down' approach, which starts from a principle, or ideal, without taking into account the existing institutional forms. The advantage of pursuing an institutionalist approach is that there are certain features of the world today that are unlikely to change. For instance, the fact that the world is divided into nation-states may not be something that people would aim for in a non-institutional approach, but it is something that people have to accept when determining principles of

[39]See Stephen Gardiner's commentary on institutional inadequacy (Gardiner 2011, p. 28).

[40]Chayes and Chayes 1995.

[41]For more on this point, see Barrett 2003; Lawrence 2014.

[42]See: Blake 2001; Griffin 2008.

institutional design. But there are also downsides to this approach. People might fail to achieve the maximum attainable ideals, or they might accept practices or institutions that should be outlawed. These problems can be avoided by keeping in mind that we're only taking an institutionalist approach to policy. More radical proposals for reform should also play a role in shaping our thoughts about the ultimate ends that people should pursue.

1.6 Chapter Outline

The ultimate goal of this book is to develop a set of normative principles that guide the procedural design of the UNFCCC. In doing so, it also provides an argument for considering procedural justice in the UNFCCC, and advocates continued support for international cooperation in this forum. It does so in four steps. First, it argues that there is reasonable disagreement over the fair distribution of the costs and benefits of climate change. Second, it argues that where there is reasonable disagreement over such ends, fair procedures can provide a way of reaching agreement amongst those who disagree. Third, it then identifies several principles of procedural justice and evaluates these principles alongside other criteria to determine a set of rules for decision-making in climate institutions. Lastly, the book uses these findings to specify some pragmatic policy proposals for reforming the procedural design of the UNFCCC. The following is a brief account of the chapters that make up the rest of this book.

Chapters 2 and 3 make a case for fair procedures in the UNFCCC. Chapter 2 identifies and analyses the different concepts of fairness that have been put forward in the UNFCCC, arguing that each of these principles represents a 'reasonable' proposal of what's fair. In specific, it looks at several proposals that have been advocated for the fair distribution of emission rights in the UNFCCC. It does this by first developing a notion of 'reasonableness' based on liberal political theory and then comparing the most prominent principles that have been proposed in the UNFCCC against this standard. It argues that, because the disagreement in the UNFCCC represents 'reasonable disagreement', it is very unlikely that parties will arrive at a mutually acceptable outcome through deliberative debate.

Chapter 3 builds on Chap. 2 by showing that fair procedures can help actors to avoid political stalemate in situations of reasonable disagreement. It does this by arguing that a fair procedure can help actors to reach a fair compromise on a position, even if each actor holds a different position on an issue. First, it discusses the relative merits of fair procedures, arguing that, whilst fair procedures are important in themselves, there are sometimes trade-offs between designing a process that is procedurally fair and designing a process to meet other more pressing ends. Second, it argues that there are certain requirements for addressing climate change including stringency, urgency, and voluntary cooperation. This means that it is important to find a way of reaching agreement in the UNFCCC even when there is reasonable disagreement over some of the ends that it should bring about.

Third, given these specific characteristics and requirements, and given the existence of reasonable disagreement, I show that fair procedures theoretically provide a way of reaching agreement in the UNFCCC even when there is reasonable disagreement over the ends that it should pursue.

Having shown why procedural values are important for the UNFCCC, Chaps. 4, 5, 6, and 7 then identify what procedural justice requires for the UNFCCC. Chapter 4 identifies who should participate in the decision-making processes of the UNFCCC. Procedural justice is often understood as requiring that all those who are affected by the outcome of a decision should have some say in the decision making process (the All Affected Principle). Yet, there are many objections to this approach, and it is not immediately apparent that this principle should be applied in climate institutions. Furthermore, increasing the number of participants in a decision is detrimental to making decisions quickly. In this chapter, I discuss the merit of the All Affected Principle and consider how fair participation can be achieved in efficient decision-making processes. I propose that decisions should include those who are coerced by a decision, and those in whose name a decision is made. I also argue that those whose interests are affected by our decisions have a right to express their interests in a debate.

The purpose of Chap. 5 is twofold. First, it explores the intrinsic value of procedural justice by arguing that there is a deep connection between procedural justice and democracy and that the same normative ideals form the basis of these two concepts. Following this, it claims that fair procedures are those that respect autonomy, equality and justification. The second aim of this chapter is to discuss what this means for equality in the decision-making processes of the UNFCCC. Asymmetries in financial resources, technical expertise and scientific information affect the ability of actors to represent themselves in decisions in the UNFCCC, highlighting that formal rights to participation in negotiations are insufficient for fairness. This chapter shows that political equality requires an equal opportunity to participate, as well as an equal playing field for making decisions.

Chapter 6 considers what's required for fair bargaining in the UNFCCC. In addition to political equality, there are also concerns that power inequalities within the decision-making processes of the UNFCCC may prohibit the 'fair terms of co-operation' between bargaining actors.[43] This is because the outcomes of negotiated agreements are partly determined by the relative bargaining power between parties. Outcomes negotiated in this way are criticised as morally unacceptable, and many call for 'responsible' or 'principled' negotiations to address this problem.[44] This chapter identifies the necessary conditions for fair bargaining in the decision-making processes of the UNFCCC. It argues that bargaining processes are fair provided that they meet some requirements of voluntariness and reciprocity.

The purpose of Chap. 7 is twofold. First, it identifies which voting mechanism should be used in the UNFCCC. Second, it specifies how votes should be weighted

---

[43]Shue 1992; Miller 2005, p. 75.

[44]Grasso 2010, p. 61; Müller 2010; Winkler and Beaumont 2010, p. 646.

in the decision-making processes of these institutions. Traditionally, decisions in the UNFCCC are made by consensus, which is criticised as promoting lowest common denominator outcomes and encouraging uncooperative behaviour by parties.[45] Given the deadlock within this institution, some propose switching to voting by majority rule.[46] Adopting majority rule would introduce the possibility of giving more votes to those that represent larger groups or to those with a greater stake in a decision. This chapter argues that decisions in the UNFCCC should be made by majority rule. It then argues that votes should be weighted according to the number of actors that each voter represents.

Chapters 2 and 3 make a case for fair procedures in the UNFCCC. Chapters 4, 5, 6, and 7 then determine the necessary rules for these procedures. Chapter 8 uses these arguments to contribute to the debate over the future of the UNFCCC. Given that a large proportion of the world's emissions are caused by only a small number of countries, some suggest that a limited agreement amongst these actors is all that is needed for climate change. In contrast, the book concludes by providing a new an innovative argument in support of the UNFCCC. Chapter 8 proposes that the procedural fairness is necessary for an effective climate change regime. Further, based on the findings of Chaps. 4, 5, 6, and 7, it argues that the UNFCCC is the only forum in which procedural fairness is possible. Whilst smaller multilateral arrangements are important measures for coordinating action on climate change, these should be pursued alongside the comprehensive approach of the UNFCCC.

References

Adger, W.N., K. Brown, et al. 2003. Governance for sustainability: Towards a 'Thick' analysis of environmental decisions. *Environment and Planning* 35: 1095–1110.

Adger, W.N., J. Paavola, et al. 2006. *Fairness in adaptation to climate change*. Cambridge, MA: MIT Press.

Albin, C. 2001. *Justice and fairness in international negotiation*. Cambridge: Cambridge University Press.

Albin, C. 2003. Getting to fairness: Negotiations over global public goods. In *Providing public goods: Managing globalization*, ed. I. Kaul, P. Conceição, K. Le Goulven, and U. Mendoza. Oxford: Oxford University Press.

Aldy, J.E., and R.N. Stavins. 2007. *Architectures for agreement*. Cambridge: Cambridge University Press.

Aldy, J.E., and R.N. Stavins. 2010. *Post-Kyoto international climate policy: Implementing architectures for agreement*. Cambridge: Cambridge University Press.

Aldy, J.E., S. Barrett, et al. 2003. Thirteen Plus One: A comparison of global climate policy architectures. *Climate Policy* 3: 373–397.

Arneson, R.J. 2004. Democracy is not intrinsically just. In *Justice and democracy*, ed. K. Dowding, R.E. Goodin, and C. Pateman. Cambridge: Cambridge University Press.

[45]Yamin and Depledge 2004, p. 443.

[46]Dimitrov 2010; Biermann et al. 2011.

Bäckstrand, K. 2006. Democratising global governance? Stakeholder democracy after the world summit on sustainable development. *European Journal of International Relations* 12(4): 467–498.

Bäckstrand, K. 2008. Accountability of networked climate governance: The rise of transnational climate partnerships. *Global Environmental Politics* 8(3): 74–102.

Bäckstrand, K. 2010. Democratizing global governance of climate change after Copenhagen. In *Oxford handbook on climate change and society*, ed. J. Dryzek, R.B. Norgaard, and D. Schlosberg. Oxford: Oxford University Press.

Bäckstrand, K., J. Khan, et al. 2010. The promise of new modes of environmental governance. In *Environmental politics and deliberative democracy: Examining the promise of new modes of governance*, ed. K. Bäckstrand. Cheltenham: Edward Elgar.

Banuri, T., K. Goran-Maler, et al. 1995. Equity and social considerations. In *Economic and social dimensions of climate change*, Contribution of working group III to the second assessment report of the Intergovernmental Panel on Climate Change, ed. J.P. Bruce, L. Hoesung, and E. Haites. Cambridge: Cambridge University Press.

Barrett, S. 2003. *Environment and statecraft*. Oxford: Oxford University Press.

Barrett, S., and R.N. Stavins. 2003. Increasing participation and compliance in international climate change agreement. *International Environmental Agreements: Politics, Law and Economics* 3: 349–376.

Bashmakov, I., C. Jepma, et al. 2001. Policies, measures, and instruments. In *Climate change 2001: Mitigation*, Contribution of working group III to the third assessment report of the Intergovernmental Panel on Climate Change. Cambridge: Cambridge University Press.

Beitz, C.R. 1989. *Political equality: An essay in democratic theory*. Princeton: Princeton University Press.

Benhabib, S. 1996. *Democracy and difference: Contesting the boundaries of the political*. Princeton: Princeton University Press.

Besson, S. 2009. Institutionalizing global Demoi-cracy. In *Legitimacy, justice and public international law*, ed. L.H. Meyer. Cambridge: Cambridge University Press.

Biermann, F. 2010. Beyond the intergovernmental regime: Recent trends in global carbon governance. *Current Opinion in Environmental Sustainability* 2: 284–288.

Biermann, F., P. Pattberg, H. van Asselt, and F. Zelli. 2009. The fragmentation of global governance architectures: A framework for analysis. *Global Environmental Politics* 4: 14–40.

Biermann, F., P. Pattberg, et al. 2010. *Global climate governance beyond 2012: Architecture, agency and adaptation*. Cambridge: Cambridge University Press.

Biermann, F., K. Abbott et al. 2011. *Transforming governance and institutions for global sustainability: Key insights from the earth system governance project*. Earth System Governance Working Paper No. 17. Lund/Amsterdam: Earth System Governance Project.

Blake, M. 2001. Distributive justice, state coercion, and autonomy. *Philosophy & Public Affairs* 30(3): 257–296.

Bodansky, D. 1993. The United Nations framework convention on climate change: A commentary. *Yale Journal of International Law* 18(2): 451–558.

Bodansky, D. 2001. The history of the global climate change regime. In *International relations and global climate change*, ed. U. Luterbacher and D.F. Sprinz. Cambridge, MA/London: MIT Press.

Bodansky, D. 2007. Legitimacy. In *The Oxford handbook of international environmental law*, ed. D. Bodansky, J. Brunnée, and E. Hey. Oxford/New York: Oxford University Press.

Bodansky, D., and S. Chou. 2004. *International climate efforts beyond 2012: A survey of approaches*. Washington, DC: Pew Center on Global Climate Change.

Bosetti, V., and J. Frankel. 2012. Politically feasible emissions targets to attain 460 ppm CO_2 concentrations. *Review of Environmental Economics and Policy* 6(1): 86–109.

Buchanan, A.E. 2004. *Justice, legitimacy, and self-determination: Moral foundations for international law*. Oxford: Oxford University Press.

Buchanan, A.E., and R.O. Keohane. 2006. The legitimacy of global governance institutions. *Ethics and International Affairs* 20(4): 412.

Bulkeley, H., and P. Newell. 2010. *Governing climate change*. Oxon/New York: Routledge.
Caney, S. 2006. Cosmopolitan justice, rights and global climate change. *Canadian Journal of Law and Jurisprudence* 19: 255–278.
Caney, S. 2009. Climate change, human rights and moral thresholds. In *Human rights and climate change*, ed. S. Humphreys and M. Robinson. Cambridge: Cambridge University Press.
Chasek, P., and L. Rajamani. 2003. Steps toward enhanced parity: Negotiating capacities and strategies of developing countries. In *Providing public goods: Managing globalization*, ed. I. Kaul, P. Conceição, K. Goulven, and F. Mendoza. Oxford: Oxford University Press.
Chayes, A., and A.H. Chayes. 1995. *The New Sovereignty: Compliance with international regulatory agreements*. Cambridge, MA/London: Harvard University Press.
Chimni, B.S. 2004. International institutions today: An imperial global state in the making. *The European Journal of International Law* 15(1): 1–37.
Cohen, J. 1989. Deliberation and democratic legitimacy. In *The good polity: Normative analysis of the state*, ed. A. Hamlin and P. Petit. New York: Blackwell.
Cohen, J. 1997. Procedure and substance in deliberative democracy. In *Deliberative democracy: Essays on reason and politics*, ed. J. Bohman and W. Rehg. Cambridge, MA/London: MIT Press.
Corbera, E., and H. Schroeder. 2011. Governing and implementing REDD+. *Environmental Science & Policy* 14(2): 89–99.
Daniels, N. 1996. *Justice and justification: Reflective equilibrium in theory and practice*. Cambridge/New York: Cambridge University Press.
Depledge, J. 2005. *The organization of global negotiations: Constructing the climate change regime*. London: Earthscan.
Depledge, J., and F. Yamin. 2009. The global climate change regime: A defence. In *The economics and politics of climate change*, ed. D. Helm and C. Hepburn. Oxford: Oxford University Press.
Dimitrov, R.S. 2010. Inside UN climate change negotiations: The Copenhagen conference. *Review of Policy Research* 27(6): 795–821.
Dingwerth, K. 2005. The democratic legitimacy of public-private rule making: What can we learn from the world commission on dams? *Global Governance* 11(1): 65–83.
Dryzek, J., and S.J. Niemeyer. 2006. Reconciling pluralism and consensus as political ideals. *American Journal of Political Science* 50(3): 634–649.
Dryzek, J., and H. Stevenson. 2011. Global democracy and earth system governance. *Ecological Economics* 70(11): 1865–1874.
Dryzek, J., and H. Stevenson. 2012a. The discursive democratization of global climate governance. *Environmental Politics* 21(2): 189–210.
Dryzek, J., and H. Stevenson. 2012b. Legitimacy of multilateral climate governance: A deliberative democratic approach. *Critical Policy Studies* 6(1): 1–18.
Dubash, N.K. 2009. Copenhagen: Climate of mistrust. *Economic & Political Weekly* XLIV(52): 8–11.
Foti, E., L. de Silva, et al. 2008. *Voice and choice: Opening the door to environmental democracy*. Washington, DC: Word Resources Institute.
Franck, T. 1995. *Fairness in international law and institutions*. Oxford: Clarendon.
Fu-Bertaux X, and A. Ochs. 2012. *An uncertain mandate from Durban*. Worldwatch Institute. www.worldwatch.org/node/9491
Gardiner, S. 2011. *A perfect moral storm: The ethical tragedy of climate change*. New York: Oxford University Press.
Garnaut, R. 2009. *The garnaut report*. Swindon: Economic and Social Research Council.
Ghosh, A. 2010. *Making climate look like trade? Questions on incentives, flexibility and credibility*, Policy brief for centre for policy research. Chanakyapuri: Dharma Marg.
Global Commission on the Economy and Climate. 2014. *The new climate economy*. Available at: http://newclimateeconomy.net/content/global-commission.
Grasso, M. 2007. A normative ethical framework in climate change. *Climate Change* 81: 223–246.
Grasso, M. 2010. *Justice in funding adaptation under the international climate change regime*. Dordrecht/New York: Springer.

Grasso, M. and J. Timmons Roberts. 2013. A fair compromise to break the climate impasse. *Global Economy and Development at Brookings*, 2013–02.
Griffin, J.W. 2008. *On human rights*. Oxford/New York: Oxford University Press.
Gupta, S., D.A. Tirpak, et al. 2007. Policies, instruments and co-operative arrangements. In *Climate change 2007: Mitigation. Contribution of working group III to the fourth assessment report of the Intergovernmental Panel on Climate Change*, ed. B. Metz, O.R. Davidson, P.R. Bosch, R. Dave, and L.A. Meyer. Cambridge: Cambridge University Press.
Gutmann, A., and D. Thompson. 1996. *Democracy and disagreement*. Cambridge, MA: Harvard University Press.
Habegger, B. 2010. Democratic accountability of international organizations: Parliamentary control within the council of Europe and the OSCE and the prospects for the United Nations. *Cooperation and Conflict* 45(2): 186–204.
Halsnæs, K., P.R. Shukla, et al. 2007. Framing issues. In *Climate change 2007: Mitigation. Contribution of Working Group III to the Fourth Assessment Report of the Inter-governmental Panel on Climate Change*, ed. B. Metz, O.R. Davidson, P.R. Bosch, R. Dave, and L.A. Meyer. Cambridge: Cambridge University Press.
Hamlin, A., and Z. Stemplowska. 2012. Theory, ideal theory and the theory of ideals. *Political Studies Review* 10: 48–62.
Hare, W., C. Stockwell, et al. 2010. The architecture of the global climate regime: A top-down perspective. *Climate Policy* 10(6): 600–14.
Harstad, B. 2009. *Rules for negotiating and updating climate treaties*. Harvard Project on International Climate Agreements. Belfer Center for Science and International Affairs, John F. Kennedy School of Government.
Harstad, B. 2010. How to negotiate and update climate agreements. In *Post-Kyoto international climate policy: Implementing architectures for agreement*, ed. J.E. Aldy and R.N. Stavins. Cambridge: Cambridge University Press.
Held, D. 2004. *Global covernant*. Cambridge: Polity Press.
Höhne, N., C. Galleguillos, et al. 2002. *Evolution of commitments under the UNFCCC: Involving newly industrialized economies and developing countries*. The German Federal Environmental Agency (Umweltbundesamt).
Huang, J. 2009. A leadership of twenty (L20) within the UNFCCC: Establishing a legitimate and effective regime to improve our climate system. *Global Governance* 15(4): 435–441.
Huq, S. 2006. Equity in national adaptation programs of action (NAPAs): The case of Bangladesh. In *Fairness in adaptation to climate change*, ed. M.R. Khan, N.W. Adger, J. Paavola, S. Huq, and M.J. Mace. Cambridge, MA/London: MIT Press.
IISD. 2010. Earth negotiations bulletin: Summary of the cancun climate change conference. *Earth Negotiations Bulletin* 12(498): 1–30.
IPCC. 1990. *IPCC special report on participation of developing countries*. Geneva: Intergovernmental Panel on Climate Change.
IPCC. 2001. Third assessment report: Climate change 2001. In *Contribution of working groups I, II and III to the third assessment report of the Intergovernmental Panel on Climate Change*, ed. R.T. Watson and The Core Writing Team. Cambridge, UK/New York: Cambridge University Press.
IPCC. 2007. Fourth assessment report: Climate change 2007. In *Contribution of Working Groups I, II and III to the Fourth Assessment Report of the Intergovernmental Panel on Climate Change*, ed. Core Writing Team, R.K. Pachauri and A. Reisinger. Geneva: IPCC.
IPCC. 2012. *Managing the risks of extreme events and disasters to advance climate change adaptation*, Special report of the Intergovernmental Panel on Climate Change. Cambridge: Cambridge University Press.
IPCC. 2013. Climate change 2013: The physical science basis. In *Contribution of working group I to the fifth assessment report of the Intergovernmental Panel on Climate Change*, ed. T.F. Stocker, D. Qin, G.-K. Plattner, M. Tignor, S.K. Allen, J. Boschung, A. Nauels, Y. Xia, V. Bex and P.M, 1535 pp. Midgley. Cambridge University Press, Cambridge, UK/New York.

IPCC WGIII. 2014. Technical summary. In *Climate change 2014: Mitigation of climate change. Contribution of working group III to the fifth assessment report of the Intergovernmental Panel on Climate Change*. Cambridge, UK/New York: Cambridge University Press.

Jagers, S.C., and J. Stripple. 2003. Climate governance beyond the state. *Global Governance* 9(3): 385–99.

Karlsson-Vinkhuyzen, S.I., and J. McGee. 2013. Legitimacy in an era of fragmentation: The case of global climate governance. *Global Environmental Politics* 13(3): 56–78.

Keohane, R.O. 2003. Global governance and democratic accountability. In *Taming globalization*, ed. D. Held and M. Koenig-Archibugi. Cambridge: Polity Press.

Keohane, R.O. 2006. Accountability in world politics. *Scandanavian Political Studies* 29(2): 75–87.

King, D., K. Richards, et al. 2011. International climate change negotiations: Key lessons and next steps. In *Smith school of enterprise and the environment*. Oxford: The University of Oxford.

Kingsbury, B., N. Krisch, et al. 2005. The emergence of global administrative law. *Law & Contemporary Problems* 68(3&4): 15–61.

Klinsky, S., and H. Dowlatabadi. 2009. Conceptualizations of justice in climate policy. *Climate Policy* 9: 88–108.

Krasner, S.D. 1982. Structural causes and regime consequences: Regimes as intervening variables. *International Organization* 36: 185–205.

Lawrence, P. 2014. *Justice for future generations*. Cheltenham: Edward Elgar.

Levi, M.A., and K. Michonski. 2010. *Harnessing international institutions to address climate change*, Council of foreign relations working paper. New York: Council of Foreign Relations.

Lomborg, B. 2001. *The sceptical environmentalist*. Cambridge, UK: Cambridge University Press.

Mason, A. 2004. Just constraints. *British Journal of Political Science* 34: 251–268.

McDermott, D. 2008. Analytical political philosophy. In *Political theory: Methods and approaches*, ed. D. Leopold and M. Stears. Oxford/New York: Oxford University Press.

MEF. 2013. *Major economies forum on energy and climate*. http://www.majoreconomiesforum.org/.

Miller, D. 2005. Against global egalitarianism. *The Journal of Ethics* 9: 55–79.

Miller, D. 2008. Global justice and climate change: How shall responsibilities be distributed? In *Tanner lectures on human values*. Delivered at Tsinghua University, Beijing, March 24–25, 2008.

Müller, B. 2010. Copenhagen 2009; failure of final wake-up call for our leaders. *Oxford Institute for Energy Studies* EV49.

Nordhaus, W.D. 2009 Economic issues in a designing a global agreement on global warming. Keynote address prepared for: *Climate change: Global risks, challenges, and decisions*. Copenhagen, Denmark, March 10–12, 2009. http://nordhaus.econ.yale.edu/documents/Copenhagen_052909.pdf.

Nye, J.S. 2001. Globalization's democratic deficit: How to make international institutions more accountable. *Foreign Affairs* 80(4): 2–6.

OECD. 2012. *OECD environmental outlook to 2050: The consequences of inaction*. OECD Publishing. http://dx.doi.org/10.1787/9789264122246-en.

OHCHR. 2009. *Report of the office of the United Nations high commissioner for human rights on the relationship between climate change and human rights*. The Office of the United Nations High Commissioner for Human Rights and the Office of the High Commissioner and the Secretary-General. http://www.ohchr.org/Documents/Press/AnalyticalStudy.pdf.

OHCHR and UNEP. 2012. *Human rights and the environment, Rio + 20*. Joint Report United Nations Human Rights Office of the High Commissioner and UNEP. OHCHR/UNEP.

Paavola, J. 2005. Seeking justice: International environmental governance and climate change. *Globalizations* 2(3): 309–322.

Pattberg, P., and J. Stripple. 2008. Beyond the public and private divide: Remapping transnational climate governance in the 21st century. *International Environmental Agreements* 8(4): 367–388.

Philibert, C., and J. Pershing. 2001. Considering the options: Climate targets for all countries. *Climate Policy* 1: 211–227.

Prins, G. and S. Rayner (2007a). *The wrong trousers: Radically rethinking climate policy*. Joint discussion paper of the James Martin Institute for Science and Civilization, University of Oxford and the MacKinder Centre for the Study of Long-Wave Events, London School of Economics.

Prins, G., and S. Rayner. 2007b. Time to ditch Kyoto. *Nature* 449: 973–975.

Rajamani, L. 2000. The principle of common but differentiated responsibility and the balance of commitments under the climate regime. *Review of European Community & International Environmental Law* 9(2): 120–131.

Rawls, J. 1971. *A theory of justice*. Cambridge, MA: Harvard University Press.

Rayner, S. and M. Thompson. 1998. Cultural discourses. In *Human choice and climate change*, ed. S. Rayner and E. Malone. Columbus: Battelle Press.

Scharpf, F. 1999. *Governing in Europe: Effective and democratic?* Oxford: Oxford University Press.

Schelling, T. 1997. The cost of combating global warming: Facing the trade offs. *Foreign Affaris*. 76: 8–14.

Schlosberg, D. 1999. *Environmental justice and the new pluralism: The challenge of difference for environmentalism*. Oxford: Oxford University Press.

Shue, H. 1992. The unavoidability of justice. In *The international politics of the environment: Actors, interests and institutions*, ed. A. Hurrell and B. Kingsbury. Oxford: Clarendon.

Stavins, R., et al. 2014. International cooperation: Agreements and instruments. In *Climate change 2014: Mitigation of Climate Change. Contribution of Working Group III to the Fifth Assessment Report of the Intergovernmental Panel on Climate change*, ed. O. Edenhofer et al. Cambridge/New York: Cambridge University Press.

Stern, N.H. 2007. *The economics of climate change: The stern review*. Cambridge: Cambridge University Press.

Stevenson, H. 2011. *Representing "The Peoples"? Post-neoliberal states in the international climate negotiations*. Centre for Deliberative Democracy & Global Governance. Working Paper (2011/1).

Swift, A. 2008. The value of philosophy in nonideal circumstances. *Social Theory and Practice* 34(3): 363–87.

Torvanger, A., and L. Ringius. 2002. Burden differentiation: Criteria for evaluation of burden-sharing rules in international climate policy. *International Environmental Agreements: Politics; Law and Economics* 2(3): 221–235.

Toth, F.L. 1999. *Fair weather? Equity concerns in climate change*. London: Earthscan.

Toth, F.L., M.J. Mwandosya, et al. 2001. Decision making frameworks. *Climate change 2001: Mitigation*. Contribution of Working Group III to the Third Assessment Report of the Intergovernmental Panel on Climate Change. Cambridge: Cambridge University Press.

UN. 1969. Vienna convention on the law of treaties. *United Nations Treaty Series* 1155: 331.

UN. 1998. Convention on access to information, public participation in decision-making and access to justice in environmental matters (The Aarhus convention). *United Nations* 2161: 447.

UNFCCC. 1992. *United Nations framework convention on climate change*. Convention: Text.

UNFCCC. 1996. *Oraganizational matters: Adoption of rules of procedure*. FCCC/CP/1996/2. Conference of the parties second session, Geneva, July 8–19, 1996

UNFCCC. 1997. *Kyoto protocol to the United Nations framework convention on climate change*. Bonn: United Nations Framework Convention on Climate Change.

UNFCCC. 2011. *The Durban platform*. Draft decision -/CP.17 Establishment of an Ad Hoc Working Group on the Durban Platform for Enhanced Action.

UNFCCC. 2012. Report of the conference of the parties on its eighteenth session, held in Doha from 26 November to 8 December 2012. UNFCCC/CP/2012/8.

UNFCCC. 2013 Report of the conference of the parties on its nineteenth session, held in Warsaw from 11 to 23 November 2013.

UNFCCC. 2014. *Lima call for climate action*. United Nations Framework Convention on Climate Change.

Victor, D. 2006. Toward effective international cooperation on climate change: Numbers, interests and institutions. *Global Environmental Politics* 6(3): 90–103.

Vihma, A., and K. Kulovesi. 2012. *Strengthening global climate change negotiations; Improving the efficiency of the UNFCCC process*. Nordiske Arbejdspapirer Nordic Working Papers. Nordic Council of Ministers.

Waldron, J. 1999. *Law and disagreement*. Oxford/New York: Oxford University Press.

Winkler, H., and J. Beaumont. 2010. Fair and effective multilateralism in the post-Copenhagen climate negotiations. *Climate Policy* 10(6): 638–654.

Yamano, H., et al. 2007. Atoll island vulnerability to flooding and inundation revealed by historical reconstruction: Fongafale Islet, Funafuti Atoll, Tuvalu. *Global and Planetary Change* 57: 407–416.

Yamin, F., and J. Depledge. 2004. *The international climate change regime: A guide to rules, institutions and procedures*. Cambridge: Cambridge University Press.

Ypi, L. 2010. On the confusion between ideal and non-ideal in recent debates on global justice. *Political Studies* 58: 536–555.

Zürn, M. 2000. Democratic governance beyond the nation state. *European Journal of International Relations* 6(2): 183–221.

Chapter 2
Reasonable Disagreement and Political Deadlock

2.1 Introduction

This book analyses the fairness of the decision-making procedures of the UNFCCC. Chapters 4, 5, 6, and 7 provide an account of what fair decision-making procedures would look like for the UNFCCC, as well as a comparison of these ideals against its existing procedures. Before doing this, it's worth making a case for considering procedural fairness in the UNFCCC in the first place. This is the purpose of Chaps. 2 and 3. In this chapter, I argue that there is reasonable disagreement over some of the positions that are advocated in the UNFCCC. This is important because it means that deliberation is unlikely to bring about agreement in the UNFCCC anytime soon. Having done this, in Chap. 3 I then go on to argue that, where there is reasonable disagreement over such ends, fair procedures can help actors to reach agreement. Further, I argue that, because of the importance and urgency of avoiding dangerous climate change, the fact that there is reasonable disagreement in the UNFCCC provides a strong case for procedural fairness.

Following this, this chapter sets out to show that there is reasonable disagreement over some of the positions that are advocated in the UNFCCC. To substantiate this claim, I review several proposals for a specific area of discussion in the UNFCCC, namely: the fair distribution of greenhouse gas emission rights. I argue that, not only is there disagreement about the correct moral criteria for the fair distribution of emission rights, but also that this disagreement is *reasonable*. It is *reasonable* because it reflects an inability on the part of actors to reach agreement despite acting in good faith and seeking to arrive at a just outcome. Reasonable actors cannot reject proposals advocated under these conditions as reasonable interpretations of what a fair distribution is, even if they disagree with the correctness of the proposal itself. The implication of reasonable disagreement is that consensual agreement on the fair terms of cooperation is unlikely to come about even if actors are actively seeking this end. Further, the fact that this disagreement is *reasonable* presents an opportunity for achieving a mutually acceptable outcome despite initial disagreement about

L. Tomlinson, *Procedural Justice in the United Nations Framework Convention on Climate Change*, DOI 10.1007/978-3-319-17184-5_2

which substantive ends to pursue. The fact of reasonable disagreement combined with other properties of climate change makes the case for considering procedural fairness in the UNFCCC.

2.2 Reasonable Disagreement

One area of the UNFCCC in which there has been significant disagreement about substantive ends in both theory and practice is the fair distribution of emission rights. The ultimate aim of the UNFCCC is to stabilise atmospheric concentrations of greenhouse gases at a level that prevents dangerous climate change.[1] Greenhouse gas emissions are cumulative, so it is typically thought that achieving this goal requires setting some total global emissions budget and distributing this budget to states in the form of 'emission rights'. Questions then arise over *how* this budget should be distributed. Whilst Article 3 of the UNFCCC states that parties should act 'on the basis of equity and in accordance with their common but differentiated responsibilities and respective capacities', it is not clear what this actually means in practice, and there has been a great deal of debate on how emission rights should be distributed since the formation of the UNFCCC, both in academic circles and in COP negotiations. Disagreement over the issue is considered at least partly responsible for the current stalemate in the UNFCCC.

Whilst there is clearly disagreement over this issue, the purpose of this chapter is to show that there is *reasonable* disagreement. I do this in Sect. 2.3, where I analyse the various principles that have been proposed in the UNFCCC. But before analysing whether there is *reasonable* disagreement in the UNFCCC, it is necessary to explain what this actually means.

2.2.1 *Disagreement*

Of course, before clarifying what I mean by *reasonable* disagreement, it helps to say something about disagreement first. The notion of disagreement suggests that actors hold different opinions, ideas, or claims about an issue at hand.[2] It is distinct from the idea of difference, which merely implies that actors do not have the same characteristics. Rather, disagreement implies that there is an absence of consensus on an issue and that each party thinks that the other's position is wrong. Because this chapter is ultimately concerned with disagreement in the UNFCCC, I want to focus on a specific area of disagreement, namely: disagreement over the fair terms of cooperative action. That is, this chapter considers situations where there

[1]UNFCCC Convention text article 2.

[2]For a comprehensive account of disagreement, see: Besson 2005.

is a need for cooperation on matters of common concern, and there is a common understanding that the terms of this cooperation should be fair. At the same time, there is disagreement about what these terms should be.

This reflects the situation in the UNFCCC itself, where climate change requires collective action, and all parties to the convention have agreed to participate in the cooperative project. At the same time, there is disagreement about the exact terms of cooperation. There are different types of disagreement over this issue, and the likelihood of moving towards agreement depends in part on what type of disagreement actors are faced with. For this reason it is worth distinguishing between these different types of disagreement that might arise in relation to the terms of a cooperative arrangement.

(1) **Epistemic disagreement:**
Actors disagree about whether or not something is correct.

Epistemic disagreement can be disaggregated into two types:

(i) **Empirical disagreement**
People disagree about empirical facts.
(ii) **Normative disagreement**
People disagree about normative values.

(2) **Interpretative disagreement**
Actors agree that something is true, but they disagree on how best to interpret its meaning.
(3) **Weighting disagreement**
Actors agree that something is true, and how best to interpret it, but they disagree on the relative weight or importance that it should be given.
(4) **Methodological disagreement**
Actors disagree on the methodology that they should use to make a judgement on an issue.

This typology is not exhaustive; there are many other kinds of disagreement about collective action. The reason for highlighting these specific types of disagreement is that they are the most relevant forms of disagreement that arise over the fair distribution of emission rights.

The important point to draw from this typology is that some types of disagreement are likely to be more intractable than others. For instance, some epistemic disagreement is just a matter of fact; disagreement can be resolved by providing the correct information about empirical facts, or by ensuring that actors cooperate in good faith. Other types of disagreement are much more intractable.[3] Matters of interpretative or weighting disagreement are likely to rest on differences that cannot be resolved by providing correct information, or by cooperating in a fair

[3]Samantha Besson draws on empirical work to show how people tend to agree on some concepts more than others (Besson 2005, p. 156).

way. The full implications of this point are to follow. For the moment, it is only necessary to point out that different forms of disagreement are easier, or harder to resolve.

This typology does not explain how disagreement can come about. There are many reasons why actors disagree about collective decisions. As Amy Gutmann and Dennis Thompson point out, actors might disagree simply because they adopt self-interested positions in an argument, which are likely to be different for different actors (Gutmann and Thompson 1996, p. 18). This explains why actors disagree in certain situations. But here I'm primarily concerned with the notion of *reasonable* disagreement, rather than disagreement more generally. As I alluded to earlier, the fact that there is *reasonable* disagreement in the UNFCCC necessitates attention to procedural fairness. For this reason, I move on to discuss the concept of reasonableness in the following section, before moving to a discussion of how reasonable disagreement can arise in the section following that.

2.2.2 Reasonableness

Reasonableness is a property that governs how actors behave, and subsequently the types of claims that they are willing to make and accept in a fair procedure.[4] A reasonable proposal is not simply a matter of self-interest, nor is it wholly altruistic. Rather, a reasonable actor is one that conforms to certain requirements, which fall into two categories: sound reasoning, and reciprocity.

First, reasonable actors are those who act under certain conditions of reasoning. That is, reasonable actors act in accordance with principles that govern the competency of the actor's reasoning and judgement. These conditions require, for example: that an actor reasons in a consistent way, using methodologies that are commonly understood. This is the first requirement of reasonableness.

Requirement 1 Reasonable actors act in accordance with certain intellectual conditions governing sound and competent reasoning

The first requirement of reasonableness implies that actors act with sound reasoning, in the sense that they do not employ unsound, or incapable reasoning, and there are no obvious logical errors or flaws. But reasonableness concerns more than just sound reasoning. As Thomas Scanlon argues, one can reason competently, whilst also behaving very unreasonably (Scanlon 1998, p. 192–3).[5] In a bargaining situation, a rational actor may adopt hard bargaining tactics, mislead others, or exploit those who are in weaker positions. Whilst such behaviour may meet the

[4]This section draws from the discussions of reasonableness given by John Rawls, Samantha Besson, and Christopher McMahon (see: Rawls 1993, p. 49; Besson 2005, p. 91; McMahon 2009).

[5]Scanlon contrasts 'rationality' with 'reasonableness'. Rationality is a term that has many different interpretations in the literature. I therefore avoid using it here.

first requirement of reasonableness defined above, it is clearly not reasonable in a broader sense of the term. Rather, reasonableness also requires some normative constraints on the behaviour of an actor.

In his discussion of reasonableness, John Rawls provides a useful starting point for thinking about this additional requirement. In *Political Liberalism*, Rawls provides two conditions for reasonableness that incorporate this normative concern. The second of Rawls' conditions relates to the 'burdens of judgement', which I discuss in the following section. For the moment, I am concerned with the first of Rawls' conditions, which states that:

> 'Persons are reasonable in one basic aspect when, among equals say, they are ready to propose principles and standards as fair terms of cooperation and to abide by them willingly, given the assurance that others will likewise do so'. (Rawls 1993, p. 49)

Reasonableness not only implies that people act under certain conditions of competent reasoning. It also implies that people constrain the types of claims that they make according to some normative requirement. Rawls provides this necessary requirement by providing a normative commitment in his first condition of reasonableness. Rawls' first condition therefore provides the basis for our own second requirement of reasonableness:

Requirement 2 Reasonable actors are those that propose claims that represent fair terms of cooperation, under conditions of truth and good-faith reasoning

Reasonable actors do not pursue merely self-interested ends. Reasonable actors make claims that appeal to a sense of fair cooperation.[6] Now, the important upshot of all this is that other reasonable actors cannot dismiss claims that are made under these conditions as unreasonable. That is, if a reasonable actor puts forward a claim then other actors cannot reject this claim *as a reasonable proposal* for the fair terms of cooperation. To be sure, other reasonable actors *can* reject this claim as an incorrect proposal in their own opinion. But the fact that the claim is made under conditions of reasonableness means that it should be respected as a reasonable proposal for the fair terms of cooperation. This is because reasonable actors are willing to engage with each other in a fair way for the purposes of identifying fair terms of cooperation. They are therefore willing to recognise the reasonableness of another actor's claims and treat them appropriately. The requirements of reasonableness put the necessary constraints on the types of claims that one can make so that these claims are justifiable to other actors who are also seeking fair terms of cooperation. Reasonableness therefore implies a willingness to accept the views of others as well as to temper one's own claims.

[6]Christopher McMahon argues that this second requirement concerns the concessions that an actor is willing to make for the sake of fair cooperation. As such, in McMahon's terms, reasonableness is confined to cooperative contexts. But whilst reasonableness does govern the types of concession one is willing to make in a cooperative context, it also concerns the types of claim that one makes prior to any notion of compromise or concession. Here I focus on the latter claim, although I develop the fairness of compromises later in chapter six (McMahon 2009, p. 19).

These two requirements form the basis of our understanding of reasonableness. This represents a starting point for considering what a reasonable claim is, and what people should expect from reasonable actors.

2.2.3 *Reasonable Disagreement*

Having specified a typology of disagreement and outlined some properties of reasonableness I now turn my attention to the concept of reasonable disagreement. Put simply, reasonable disagreement is disagreement that exists among reasonable actors. It represents a failure on the part of actors to agree on an issue despite the fact that they are acting under the conditions of reasonableness outlined above. Following the discussion so far, reasonable disagreement means two things: (i) reasonable actors will not converge on any one answer, and (ii) reasonable actors should accept that other reasonable actors hold different views and that those views should not be dismissed as unreasonable.[7] The important point to note here is that, because each view can be reasonably upheld, such views cannot be dismissed as unreasonable. Whilst each actor can reasonably deny that the answers that others endorse are correct, these actors cannot dismiss these claims as unreasonable. Therefore, someone can reasonably reject a particular proposal as the right answer, whilst also accepting that it is not an unreasonable position to adopt. This means that no single view is likely to enjoy the full support of reasonable people. This is obviously a problem in situations that require the consensual cooperation of an entire group of actors.

The fact that disagreement is *reasonable* is important. Reasonable disagreement makes it more likely that actors will be unable to reach agreement through continued deliberation. Even if actors act cooperatively and engage with each other fairly, they are unlikely to come to an agreement about which substantive ends they should collectively pursue. This is because reasonable disagreement is not disagreement that arises out of mistake or self-interest, but rather reflects the fact that actors have different worldviews and experiences, which is an issue that I expand and clarify in the next section.

But there's a further reason that reasonable disagreement is important. If we are merely faced with a situation of disagreement among actors, then it should still be possible to specify some substantive outcome that a given procedure should aim to achieve. Whilst different actors may disagree because they hold different views that are based on mistaken reasoning, or incorrect evidence, this doesn't mean that it isn't possible to determine what the correct outcome actually is and then

[7] As Samantha Besson suggests, one might realise that another actor's views are reasonable when understood against their own background of existing beliefs, whilst still holding them incorrect from one's own viewpoint (Besson 2005, p. 114).

convince the disagreeing parties that they should endorse this outcome. Of course, if it's necessary to gain each actor's voluntary cooperation on an issue then it is still necessary to persuade the disagreeing parties to accept this procedure. If each actor's interests persistently clash then there might be persistent disagreement here. But whilst achieving agreement may be difficult in such cases, it is not a wholly impossible thing to hope for. What is more problematic is the fact that it isn't possible to specify a correct outcome in cases of reasonable disagreement. This is because reasonable disagreement reflects difference that should be respected, rather than dismissed as erroneous.

The upshot of this is that, if there is reasonable disagreement, then agents are unlikely to reach agreement on the substantive merits of particular proposals for the fair terms of cooperation. The fact that disagreement is reasonable means that actors are unlikely to reach an agreement even when they argue in good faith and for the public good.

Of course, there are other important implications of reasonable disagreement that relate to how people should think about fair procedures. There are also unresolved questions concerning why actors should act reasonably in the first place and what should be done about those who behave unreasonably. There have been many instances in COP negotiations where parties have acted unreasonably, and there are questions about the extent to which concessions should be made to these actors. But this is to get ahead of my current argument. I discuss the full implications of reasonable disagreement in the latter parts of this section. Here, the most pressing question to ask is why actors should treat reasonable claims with respect. I answer this question by considering how reasonable actors come to hold different views in the first place.

2.2.4 *The Burdens of Judgement*

If reasonableness requires that certain conditions of reasoning are met, then how can reasonable actors come to disagree in the first place? After all, reasonable actors employ sound reasoning and engage in cooperative behaviour, but reasonable disagreement presumes that reasonable actors will disagree regardless. So how can reasonable actors fail to reach agreement on an issue?

As I pointed out earlier, actors can disagree for many reasons. But I'm concerned with why *reasonable* actors disagree. Here I draw on John Rawls' discussion of the 'burdens of judgement', which he uses to explain how reasonable people can come to have different opinions on a matter (Rawls 1993, p. 54–8). Rawls argues that these are 'the hazards involved in the correct (and conscientious) exercise of our powers of reason and judgement in the ordinary course of political life' (Rawls 1993, p. 55–6). Whilst Rawls uses the burdens of judgement to explain disagreement about the comprehensive philosophical doctrines, these can also be used to explain disagreement about political ends (Waldron 1999, p. 112). The burdens of judgement are:

(a) Evidence can be conflicting and complex, and thus hard to assess and evaluate
(b) People can disagree about the weighting that they should give to different issues
(c) Concepts are vague and subject to hard cases, which means that people often have to rely on judgement and interpretation
(d) People's moral and political values are shaped by their different experiences
(e) Different normative considerations can produce difficulties in making an overall assessment
(f) Because no system of social institutions can incorporate the full range of moral and political values, people are forced to prioritise and restrict these

According to Rawls, this means that actors are likely to disagree even when they act in a reasonable way.[8] The fact that different positions can be reasonably held means that, if there is a large number of heterogeneous parties, there is unlikely to be a single option that all parties immediately agree is correct. Whilst unreasonable disagreement can be resolved by exposing self-interested or erroneous views, reasonable disagreement can persist despite continued deliberation. This suggests that reasonable disagreement is more intractable than other forms of disagreement, which would be more easily resolvable under deliberative discussion.

But this is not the main issue here. The important point to take from Rawls' discussion is that some issues require a judgement or interpretation on the part of the actor. Other issues require judgements about the values that people hold and how to prioritise them. The way that each actor judges, or interprets an issue depends on his or her previous experiences and worldviews. Given that each actor's experiences and worldviews are different, each actor judges or interprets issues differently. This means that reasonable actors can come to hold different opinions of the same thing. But this difference is not just a question of mistake or error on the part of an actor. Nor is it a case of differential self-interest. Rather, it is the fact that actors are different that leads them to disagree, even if they are acting cooperatively. This brings us to Rawls' second requirement of reasonableness, which states that not only do the burdens of judgement explain how reasonable disagreement can arise, but also that, under his account of reasonableness, reasonable people will recognise the burdens of judgement as reasons for accepting the reasonableness of others' views (Rawls 1993, p. 54). The burdens of judgement therefore play a secondary role, not only explaining how reasonable disagreement can arise, but also providing reasons for why people should respect and acknowledge the reasonableness of other people's views, even if they disagree.

[8]Rawls 1993, p. 58.

2.2.5 *Summary*

Drawing on what has been argued above, it is worth clarifying the various points made so far. Here, I define reasonableness and its associated concepts as follows:

(i) *Reasonableness*: is a quality that relates to the way that actors behave. It concerns and the types of claim that an actor is willing to accept as well as the claims that an actor should make.
(ii) *Reasonable actors*: are those who act in accordance with the requirements of reasonableness. These requirements concern reasoning under conditions of good faith and competency.
(iii) *Reasonable claims*: are claims that are made by reasonable actors. Reasonable actors cannot reject reasonable claims as unreasonable interpretations of the fair terms of cooperation.
(iv) *Reasonable disagreement*: exists when reasonable actors hold different views or opinions about a matter. It suggests that consensus is unlikely to provide a solution to reaching agreement.

The result of this is that if a claim is reasonable, then another reasonable actor cannot reject it as unreasonable. This is not to say that a reasonable actor cannot reject a reasonable claim as wrong, or incorrect. A reasonable actor may hold his or her own views about what is right in a certain context, and reject other reasonable proposals as incorrect. But reasonable actors recognise the status of reasonable claims and treat them with due respect accordingly. This means accepting that other reasonable people hold different views and not dismissing these views as unreasonable. Following this, the next section shows that there is reasonable disagreement over the distribution of emission rights by showing that there are different reasonable proposals for the fair distribution of this good. Because each of these proposals is reasonable, reasonable actors cannot dismiss any of them as unreasonable.

2.3 The Fair Distribution of Emission Rights

Having defined reasonableness I now examine some of the proposals for how people should share greenhouse gas emission rights and show that there is reasonable disagreement concerning the allocation of this good. To this end, this section first introduces the concept of allocation rules before considering the reasonableness of the prominent arguments for the fair distribution of this good.

2.3.1 *Allocation Rules for the Distribution of Emission Rights*

Many allocation rules have been proposed in the UNFCCC for the fair distribution of emission rights, both indirectly from academic circles and political activists and directly from state delegations.[9] The Fourth Assessment Report of the IPCC defines 14 such allocation rules that are frequently proposed in the literature on this subject.[10] In this section I analyse some of the allocation rules that are proposed for the fair distribution of greenhouse gas emission rights, rather than for any other end. The most prominent of these rules are:

1. **Basic Emission Rights**
2. **Grandfathering**
3. **Equal Rights**
4. **Historical Responsibility**
5. **Equal Costs**
6. **Multi-criteria Proposals**

This is just a sample of all of the proposals that are made. But these allocation rules represent the most prominent allocation rules that are advocated in both theory and practice. Whilst similar issues would arise with other allocation rules, this chapter illustrates the existence of reasonable disagreement using just these four rules.

The remainder of this section considers each of these in turn. In doing so, it is possible to show four things. First, each allocation rule is a reasonable proposal for the fair distribution of emission rights. Following the argument in Sect. 2.2, the fact that each allocation rule is a reasonable proposal means that none of these proposals can be dismissed as an unreasonable interpretation of what fairness requires. Second, even if there is a consensus that one particular allocation rule is appropriate for the fair distribution of emission rights, the application of each rule involves choosing between variables that are unspecified by the allocation rule itself, and there are different reasonable interpretations of what these variables are. Third, even if there is agreement on these first two matters, there are different reasonable weights for the relative importance of each allocation rule in question. Fourth, I show that this is a matter of disagreement in both theory and practice, by highlighting where different principles are actually proposed in climate change negotiations. Taken together, these points show that the fair allocation of emission rights is subject to reasonable disagreement in a number of distinct ways and that there is no reason to expect that there will be agreement between reasonable persons on this matter. The fact that there are a number of different proposals that cannot be reasonably rejected means that there is unlikely to be one solution that

[9]For discussions regarding the range of allocation criteria proposed in the literature, see, for example: Grubb 1992, p. 312–4; Banuri et al. 1995; den Elzen 2002; Höhne et al. 2002; Bodansky and Chou 2004. For a normative discussion of this issue, see: Gardiner 2004, p. 583–589.

[10]These are: 'Equal per Capita Allocation', 'Contraction and Convergence', 'Basic Needs', 'Adjusted per Capita Allocation', 'Equal per Capita Emissions over Time', 'Common but Differentiated Convergence', 'Grandfathering', 'Global Preference Score Compromise', 'Historical Responsibility', Ability to Pay', 'Equal Mitigation Costs', 'Triptych', 'Multi-sector Convergence', and 'Multi-criteria' (Gupta et al. 2007, p. 770). Note that not all of these are based on equity considerations.

will command the reasonable assent of all of the participants of the UNFCCC. Given this, I go on to argue that it is essential to have a fair way of dealing with disagreement that can help generate a consensus in Chap. 3. With this in mind, I now consider the main allocation principles that are suggested.

2.3.2 *Basic Emissions Rights*

The Basic Emissions Rights proposal stipulates that there is a threshold level of well-being below which individuals are not obligated to limit their emissions.[11] This proposal is based on the premise that everyone should be allowed to achieve a basic minimum level of well-being. Given that achieving this basic level of well-being requires undertaking activities that generate greenhouse gas emissions, so the argument goes, there should be no limitations on the emissions that are brought about by these activities. According to Henry Shue, this means that each ought to be allowed at least the minimum amount of emissions necessary for a decent life.' (Shue 1995, p. 252)

The Basic Emission Rights proposal receives strong support in the literature, and almost every proposal for the fair distribution of emission rights assumes that mitigation policies should only apply to those who already enjoy a threshold level of welfare. There is a moral threshold here and people shouldn't enact policies that prevent people from meeting that threshold.

Despite the appeal of this proposal, there is at least one way in which its suitability as a principle of fairness might be reasonably disputed. The Basic Emission Rights proposal is premised on the assumption that there is a sufficient emissions budget to provide everyone with the necessary emissions to meet his or her basic needs. But because climate change is already occurring, one might reasonably argue that there is insufficient atmospheric space to exempt those emissions that are necessary for people to meet these needs. Whilst people frequently argue that the climate can tolerate a 2 °C rise in preindustrial temperatures without causing dangerous climate change, some hold that a much more stringent target is needed, even if this means preventing some from meeting their basic needs (Allen 2009). A great deal of harm is already being caused to the planet and there are many uncertainties about potential positive feedback loops, threshold limits, and tipping points in the climate system that may bring about severe and catastrophic changes to the environment. It is therefore reasonable to propose that there is an insufficient carbon budget to implement the Basic Emission Rights proposal.[12] At the same time, it is also reasonable to hold that the atmosphere can tolerate at least some limited amount of further emissions before people risk causing dangerous climate

[11] Pan 2003, p. 8; Baer and Athanasiou 2007, p. 8; Heyward 2007, p. 523.

[12] Of course, this also applies to other allocation rules. For more on the carbon budget see: Meinshausen et al. 2009; Allen et al. 2009a, b.

change, and that there is sufficient atmospheric space for providing basic emission rights so that people can meet their basic needs. That is, because the full effects of greenhouse gas emissions are unknown, there may be a sufficient carbon budget to implement the basic emission rights proposal.

Both of these positions represent different but reasonable positions about whether the Basic Emissions Rights proposal is suitable as a fair allocation rule. Since these claims are both reasonable, neither can be reasonably rejected as interpretations of what constitutes dangerous interference with the atmosphere. Disagreement arises here for two separate reasons. On the one hand, people may disagree about what atmospheric concentration of greenhouse gases constitutes dangerous climate change, which is a form of empirical disagreement (following the typology in Sect. 2.2). On the other hand, people may agree that it is desirable both to achieve a minimum level of well-being and to avoid dangerous climate change, whilst disagreeing about what should take priority here. Different reasonable positions cannot be rejected on this issue. Because this reflects the relative importance that an actor places on an issue, this represents a form of weighting disagreement.[13]

Even if there is agreement that there is sufficient atmospheric space for the Basic Emissions Rights proposal, it is not clear what the threshold level of welfare for exemption should be. The proposal for Basic Emission Rights stipulates that people should be exempt from those emissions that are necessary for achieving a basic minimum level of well-being for an individual. But determining what this level actually is involves a judgement about the minimum level of welfare that an individual should achieve. This is a matter of ongoing philosophical debate,[14] and there are different reasonable positions about what this should be. One might reasonably contend that a basic level of welfare involves having access to a wide range of life opportunities, including the right to cultural practices. Alternatively one might reasonably argue that a basic level of welfare should be as minimal as possible, incorporating only the most absolute and basic needs of an individual. The point is that each of these positions is a reasonable interpretation of what a basic level of welfare should be. Because each of these positions is reasonable, neither of them can be reasonably rejected. This means that there is reasonable disagreement about how to apply the Basic Emission Rights proposal, because there is reasonable disagreement about what level of welfare it should apply to. Since this represents a disagreement about how to interpret this allocation rule, this is a form of interpretative disagreement.

There is one further point to make regarding this proposal. Thus far I have suggested that there are different reasonable positions regarding the interpretation

---

[13]There are different reasonable positions about what to do if the atmospheric budget is insufficient for everyone to meet their basic needs. Stephen Gardiner suggests that if emissions are morally essential then they may have to be guaranteed even if this means exceeding the scientific optimum of total emissions (Gardiner 2004, p. 585). But one might reasonably claim that the stakes are too high to implement this policy.

[14]For different accounts of basic rights, including human rights, see: Shue 1996; Pogge 2005; Griffin 2008.

of this principle, owing to disagreement over what a basic level of well-being should be. But in addition to any notion of reasonable disagreement, the Basic Emission Rights proposal is an incomplete account of what a fair distribution of emission rights is because it does not state how emissions should be distributed above the minimum level needed to achieve a basic level of well-being (if indeed there is any remainder to distribute). Whilst this point does not represent an area of reasonable disagreement itself, it does indicate that this allocation rule is insufficient in determining what a fair distribution of emission rights is and that it needs to be supplemented by a further principle. There may, of course, be reasonable disagreement about what supplementary principle, or principles these should be, which is something I explore when I discuss Multi-criteria rules.[15]

2.3.3 Grandfathering

Having examined the Basic Emissions Rights proposal, I now consider a second often-proposed distributive principle: Grandfathering. Grandfathering is an allocation rule that distributes emission rights according to either existing emissions levels or a historic emission baseline.[16] Grandfathering gives more emission rights to those who generated more emissions in the past. Despite being relatively prominent in the early debates about the fair distribution of emission rights, Grandfathering has become a contentious allocation rule that has largely fallen out of the mainstream proposals. Still, it is worth considering here because, even though it is a relatively contentious proposal, different reasonable interpretations can be upheld. Further, whilst it is rarely advocated on its own, it is often advocated alongside other principles as an initial grace period that specifies a maximum rate at which actors should undertake absolute emissions reductions.[17] For example, Grandfathering is a part of the Contraction and Convergence allocation rule, which allocates emission rights so that there is a gradual convergence to equal per capita emissions at a future point in time allowing a transition period that limits the negative economic impact of reducing emission levels.[18]

There are two ways in which Grandfathering is justified as a fair allocation rule. First, Grandfathering is justified on the basis that prior use of a resource establishes a status quo right, which entitles a user to future use of this resource (Prior Use Argument).[19] That is, by virtue of using a resource in the past, an actor

[15]This is assuming that the basic rights are met and that there are remaining emission rights that can be distributed.

[16]See: Ringius et al. 2001, p. 11; Caney 2009, p. 127.

[17]For example, see: Jacoby et al. 1999.

[18]Meyer 2007, p. 11; Heyward 2007, p. 526.

[19]See the following for further explication of this rule: Banuri et al. 1995, p. 107; Meyer and Roser 2006; Bovens 2011. Meyer and Roser do not support Grandfathering, but they do argue that it

establishes a right over the future use of it. Second, Grandfathering is justified on the basis of legitimate expectations. If an investor developed greenhouse gas intensive industries at a time when climate change was not commonly known about, and these developments are now creating a lot of emissions, then there is some sense in which it is unfair if the investor is made worse off. This is because it is unfair to make an actor accountable for decisions that cause harm if these decisions were legitimate at the time that they were made.

The former of these arguments does not seem acceptable according to our everyday understanding of fairness. There are some limited cases in which people think that past use should establish a right. Disputes over land rights, in particular, often appeal to prior use. However, people don't often think that prior usage establishes a right over a resource (in fact, people often think the opposite of this), and the fact that Grandfathering typically means rewarding the better off at the expense of the worst off makes this argument unacceptable. However, the second argument for Grandfathering appears more justified. It is reasonable to propose that actors shouldn't be disadvantaged for making decisions that were legitimate at the time that they made them. Whist I do not support Grandfathering here, I do suggest that there is some sense in which some forms of this allocation rule cannot be reasonably rejected as interpretations of fairness. It seems reasonable to propose that, all else equal, some transition period should be granted to those who face very high costs in reducing greenhouse gas emissions due to investment decisions made before climate change was well known about. In this sense at least, Grandfathering cannot be rejected as an unreasonable allocation rule, due to the fact that it is based some notion of fairness.

But even if this claim is accepted, there are still problems concerning the correct application of this rule. This is because Grandfathering, as I've defined the term here, involves specifying the rate at which actors should reduce their emissions and the length of the transition period towards lower emissions levels. Different positions can be reasonably upheld regarding both of these factors because these are based on different understandings of what can be reasonably expected from actors subject to emissions reduction commitments. For instance, one might reasonably hold that actors who invested in emissions intensive industries should be given a lot of leniency in the rate at which they must reduce their emissions because one shouldn't be unfairly disadvantaged through choices that are made under legitimate expectations. At the same time, it is also reasonable to propose that these concerns should play a minimal role in our considerations of how to reduce emissions and that whilst some leniency should be given, emissions intensive industries should have to reduce their emissions at a high rate. An unreasonable proposal, on the other hand, would be one that allowed these industries to continue emitting at a high rate for a prolonged period of time, whilst others were undertaking

---

might be possible to defend this allocation rule on the basis of a historical principle of justice. This would imply that parties acquire a right to a resource through past appropriation (Meyer and Roser 2006). For more on this notion of historical justice, see: Nozick 1974.

emission reduction commitments. Therefore, even if Grandfathering is accepted as a fair allocation rule on the basis of legitimate expectations, it remains subject to reasonable disagreement because different rates and lengths of transition can be reasonably held.

2.3.4 Equal Rights

The existence of reasonable disagreement over the fair distribution of emissions rights is also shown by considering a second proposal – the Equal Rights proposal. This proposes that each is entitled to an equal share of some resource.[20] It is typically advocated on the premise that the absorptive capacity of the atmosphere is a common pool resource that all should be allowed to use to the same degree as everyone else.[21] Following this, emission rights should be distributed so that each actor receives an equal share of this resource. This is intuitively appealing. If a resource is being distributed, and no one is more deserving of it than anyone else, then it seems wholly appropriate to distribute the resource equally. But the Equal Rights proposal is not as straightforward as it at first seems. Many reasonable people may not be egalitarians, and even if one agrees that rights to use the atmosphere should be distributed equally, it is not clear what should actually be equalised here. In fact, there are various reasonable interpretations of what this should be.

The most common form of the Equal Rights proposal is the Equal per Capita Rights allocation rule. This distributes emission rights among states in proportion to the population of each state, thereby giving each person an equal quantity of emission rights.[22] An alternative form of the Equal Rights proposal is to distribute emission rights so that the welfare gains that can be drawn from using the overall emissions budget is shared equally between individuals. This is known as the Equal Welfare proposal. The distinction between the Equal per Capita Rights proposal and the Equal Welfare proposal reflects a distinction between the relative 'good' that is being distributed. In the former case, the good is an emissions right, whereas in the latter, it is the welfare that is attainable from the total emissions budget. There are different reasons for advocating each of these interpretations of the Equal Rights proposal, and both the Equal per Capita proposal and the Equal Welfare approach are reasonable proposals here.

---

[20]For discussion of the Equal Rights proposal, see: Agarwal and Narain 1991; Banuri et al. 1995, p. 106; Grubb 1995, p. 485; Singer 2002, p. 14.

[21]Singer 2002, p. 35; Traxler 2002; Gardiner 2004, p. 584.

[22]Heyward 2007, p. 521. This assumes that emission rights are distributed equally within the country to which they are allocated. Not all actors advocate this approach. For instance, David Miller argues that states should be able to distribute emission rights among their citizens as they wish (Miller 2008, p. 121).

One might reasonably advocate the Equal per Capita Rights proposal on the basis that if there is a scarce resource, this should be allocated strictly on the basis of equality. Distributing emissions rights on a per capita basis allows people to make judgements about how to use that resource according to their own preferences. It doesn't seem wholly unreasonable to claim that, if one is a strict egalitarian, and if emission rights are considered in isolation from other goods,[23] then this good should be distributed equally on a global level. Of course, there may be other reasonable people who are not egalitarians at all. But without getting into this debate, it is possible to show that there is reasonable disagreement about the relative *good* that should is subject to distribution here.

This is because the Equal per Capita Rights proposal overlooks several factors that might be considered important by reasonable actors who are considering how emission rights can be distributed fairly. For instance, the Equal per Capita Rights proposal distributes emission rights regardless of: the economic wealth of a country, the availability of alternative energy sources, and the relative energy needs that a country might have.[24] All of these properties significantly alter the benefits that an actor can get from a given amount of emission rights, and one might reasonably contend that the benefits that accrue from these rights should be equalised, rather than the right to emit itself. For this reason, some advocate the Equal Welfare approach. Because the Equal Welfare proposal takes into account the relative differences between actors, it is also a reasonable interpretation of the fair distribution of emission rights.

But the Equal Welfare proposal raises difficult questions about how much an actor's current situation is a matter of choice or circumstance. The Equal Welfare approach allocates more emission rights to those who require more of these rights to attain the same level of welfare as everyone else. However, as Ronald Dworkin has argued in his discussion of expensive tastes, it does not seem fair to allocate more resources to those who require them to achieve the same level of welfare as everyone else due to circumstances that are brought about through their own choice (Dworkin 1981, p. 228). Those who live in particularly cold climates may require additional emission rights to enjoy similar levels of welfare to those who lives in the tropics. If those who live in cold climates have had no choice over where they live then it seems fair that they should be awarded the necessary emissions to achieve the same level of welfare as those who live in warmer places. But the argument is very different if someone has chosen to live in an area where more emissions are required to achieve a given level of well-being. Whilst some have had no say over where they live, others do choose to live in certain parts of the world. This suggests that emission rights should only be given to an agent if they require more of these rights through no fault of their own. Whilst it is fair that an actor should receive

[23]This is debatable in itself. Some hold that the question of emission rights should be addressed in isolation from other issues of distributive justice, whilst others claim that it should be part of a more complete theory of distributive justice. For discussion of this point, see: Caney 2012.

[24]Risse 2008, p. 28; see also: Solomon and Ahuja 1991, p. 346; Caney 2009.

more emissions if they need them because of expensive tastes that are not of their choosing, this requirement does not hold if the actor's circumstance is a result of their own choice.

The Equal Welfare proposal therefore requires a judgement about whether an agent's situation has arisen through circumstance or choice, which is a matter on which different positions can be reasonably held. For example, if an actor chooses to live far away from her relatives, then she may need a lot of emissions to maintain close relationships with her family members. In this situation, one might reasonably argue that the agent's situation is her own making, and that if she lives far away from her family then it is her choice to do so. At the same time, others might reasonably hold that an individual's life choices are the product of many factors, and that it is naïve to suggest that people have full control over these matters. In this case, it is reasonable to argue that the agent may not be fully responsible for the fact that she lives far away from her family. Given the lack of certainty regarding the extent to which a situation is the result of choice or of circumstance, it is possible to hold different reasonable interpretations here. One might argue that this suggests that all reasonable people should adopt the Equal per Capita Rights proposal instead. But adopting the Equality of Resources proposal means that some important issues are overlooked, as I outlined in the discussion above. Whilst the Equal per Capita Rights proposal should be implemented in situations when agents are responsible for their choices, Equal Welfare is appealing when agents are not responsible for these choices, and there is reasonable disagreement about when this is the case.

The Equal Rights proposal is thus subject to reasonable disagreement on several levels. First, different reasonable positions can be held about whether equality is the correct value to apply here (normative disagreement). Second, even if there is agreement that equality is important, both the Equal per Capita Rights proposal, and the Equal Welfare proposal are reasonable interpretations of what should be equalised (interpretative disagreement about how the principle should be applied, respectively: emissions or welfare). Third, even if there is agreement about the Equal Welfare proposal, there are different reasonable ideas about how much something is a matter of choice and circumstance (empirical disagreement) and how people should think about welfare (normative disagreement) in specific cases. There is a variety of reasonable views on these issues, none of which command universal support among reasonable people and none of which can be dismissed as unreasonable. Whilst reasonable people can reject particular proposals as incorrect, they cannot dismiss these claims as unreasonable interpretations of what is fair. This means that there are a number of reasonable claims, each of which cannot be dismissed as unreasonable by reasonable actors.

2.3.5 Historical Responsibility

Of course, the Equal Rights proposal is based the idea that the fair distribution of emission rights is a resource issue, which is disputable in itself. Instead, one might reasonably hold that the problem of emission rights is one of liability. If so, one might choose to distribute emission rights according to responsibility. This is the approach of the Historical Responsibility proposal, which distributes emission rights based on an actor's previous, or historical emissions, allocating fewer emission rights to those who have emitted more in the past.[25] Historical Responsibility receives strong support in the literature[26] as well as in the UNFCCC itself under the guise of the 'Brazilian Proposal'. It is also intuitively appealing. After all, if an actor commits an act, then surely that actor should be held responsible for the consequences of that act (in some way). By allocating fewer emission rights to those who have used more of a resource in the past, the Historical Responsibility principle recognises that some have already used more of this resource than others, and thereby incorporates this notion of responsibility into the distribution of emission rights. But even if one adopts the Historical Responsibility principle as a fair allocation rule, there are different reasonable interpretations of how this principle can be achieved in practice.

One area in which this is the case concerns the identity of the agents to whom responsibility for past emissions should be attributed. Responsibility for historical emissions can be reasonably attributed to many different types of actors including: international institutions, nation-states, private companies and individuals.[27] Different positions can be reasonably held on the type of actor that is responsible for emissions according to the Historical Responsibility proposal. One reasonable position is that responsibility for creating emissions ultimately rests with the individual. But it is also reasonable to claim that many of the actions that bring about emissions are beyond of the control of individuals and that emissions should be accountable to collective actors such as states or multinational corporations. The Historical Responsibility principle requires determining who should bear responsibility here, yet there are different reasonable positions about who this should be. This is represents normative disagreement about the plausibility of different notions of collective responsibility.

A further issue relates to the criteria that people should use for holding actors responsible for their emissions. Even if there is agreement about the type of agent that is responsible for causing emissions, there are different reasonable positions regarding how responsibility for these emissions should be attributed to a certain actor. For example, it is reasonable to propose that an actor is at least partly responsible for bringing about emissions if that actor plays a causal role in a process that generates those emissions. But this raises questions about what constitutes a causal role in such a process. The current methodology of emissions accounting in the UNFCCC attributes emissions to states according to the geographical location

[25] For discussion of the relevance of historical emissions and the possible obligations emerging from them see: Rayner et al. 1999, p. 28; Shue 1999; Neumayer 2000; Gosseries 2003; Caney 2005, 2006.

[26] For example: Fuji 1990; Berk and den Elzen 1998; Neumayer 2000; La Rovere et al. 2002.

[27] Simon Caney discusses this issue (Caney 2005, 2009).

in which they are produced. That is, if a coal fired power station in country X produces emissions whilst generating power, then those emissions are accounted for as part of X's overall emissions inventory. It seems reasonable to suggest that these emissions should be attributed to X given that X is responsible for producing these emissions. However, arguments concerning how to attribute responsibility for emissions become less clear if when the products that generate emissions are consumed in a different territory to which they are produced. Imagine, for instance, that a factory in country X generates emissions whilst producing televisions that are ultimately purchased by consumers in country Y. Under the UNFCCC accounting methodology, these emissions are attributed to country X. At the same time, there is at least some sense in which the citizens of country Y are responsible for creating these emissions. Presuming that those in country Y generate a demand for televisions through their consumer choices, these citizens are also responsible for the emissions that subsequently arise. This brings up difficult questions about the relationship between causal and moral responsibility that are too great to bring up here. The purpose of this illustration is to show that there are different notions of what might reasonably be proposed as a fair interpretation of responsibility when accounting for global emissions greenhouse gas emissions.

There are many other cases where there is reasonable disagreement about causation (i.e. epistemic disagreement). Contrary to above, the factory in country X may continue to make televisions for consumers at home. In this respect, the emissions may still be brought about regardless of the choices of the citizens in country Y. As such, one might think that the consumers in country Y are not responsible for these emissions, because they are brought about regardless of their actions. It is reasonable to claim that actors shouldn't be held responsible for such indirect contributions to the processes that bring about emissions. One the other hand, one might hold that there is still some sense in which these consumers are at least partly responsible of these emissions, given that they are part of collective group of actors whose actions generate the demand for these emissions embodied goods. Both of these positions are reasonable, and represent epistemic disagreement about how to attribute responsibility appropriately. This is a common problem with causation.[28] There are different positions on this issue that cannot be dismissed as unreasonable. It is difficult to know what metric should be used to determine whether an agent is responsible, and there are different reasonable notion of what this should be.

The Historical Responsibility proposal also requires a judgement about the extent to which an actor should be held responsible for emissions that were generated before climate change became common knowledge. It is sometimes argued that an actor should not be held morally responsible for the harmful consequences of an act if the full consequences of that act were unknown at the time.[29] In relation to climate change, actors who produced emissions at a time when the negative effects

[28]For discussion, see: Mackie 1980; Hart and Honor 1985.

[29]For example: Gosseries 2005, p. 6.

of these emissions were not fully known are often said to be 'excusably ignorant' and should not be held responsible for these emissions. Consequently, advocates of the Historical Responsibility proposal often specify a date after which emitters are reasonably expected to have known about the harmful effects of their emissions and can be reasonably held responsible for them. But this raises a question about the date for excusing actors for their emissions, and there are different reasonable positions about what this should be. For instance, in the UNFCCC 1990 is frequently referred to as the latest sensible date that people could be considered excusably ignorant of the effects of producing greenhouse gas emissions.[30] It is reasonable to contend that before this time there was enough scepticism about the full effects of climate change to excuse any actor of responsibility for emissions that they produced. At the same time, the possibility that greenhouse gas emissions would cause climate change was known well before 1990.[31] It is also reasonable to claim that a much more stringent date should be set for excusing people on the ground of ignorance, because the possibility that emissions could have harmful effects were known well before this time, and past emitters should have taken this into account. Both of these proposals are reasonable claims about the extent to which actors can be reasonably expected to know about the harmful effects of climate change, representing a form of epistemic disagreement about excusable ignorance. Therefore, when applying the Historical Responsibility proposal, there are different reasonable positions about the date at which actors are excusably ignorant of the harm caused by their past emissions.

Implementing the Historical Responsibility proposal also requires a judgement about how to attribute responsibility for those emissions caused by people who are no longer alive.[32] Activities that generate emissions have been taking place since the industrial revolution. As a result, a large proportion of the total emissions already in the atmosphere were caused by actors who are no longer alive. If one advocates the Historical Responsibility proposal then it is necessary to state how to appropriately attribute responsibility for these emissions. Advocates of the Historical Responsibility proposal sometimes resolve this problem by arguing that responsibility for past emissions should be attributed to those who have benefited from these emissions (the Beneficiary Pays Principle).[33] Whilst a person cannot be held responsible for the harmful consequences of someone else's act, they still may be liable to pay compensation for something if they enjoy the corresponding benefits.[34] This suggests that those who benefit from historical emissions caused by those who are no longer alive should have fewer emission rights than those who have not.

[30]The following authors make this point: Neumayer 2000; Singer 2002; Caney 2005; Risse 2008, p. 35.

[31]The first studies suggesting that emissions might cause global warming emerged in the nineteenth century (see: Arrhenius 1896).

[32]For further discussion, see: Caney 2006.

[33]Pan 2003, p. 5; Caney 2006; Heyward 2007, p. 523.

[34]Gosseries 2005, p. 10.

Now, there is normative disagreement about whether benefitting from an act can bring about associated responsibilities to pay for the consequences of that act.[35] Whilst the Beneficiary Pays Principle is disputable in itself, even if one does endorse this principle then there are still different reasonable positions about how it is interpreted. This is because the benefits that arise from producing emissions spill over to different actors.[36] It is reasonable to claim that many of the benefits of the emissions that were created by industrialised countries have accrued to nations on a global scale. The problem is that, at this point in time, there is no way of empirically determining the extent to which different countries have benefitted from these emissions (if at all) and different reasonable claims can be made about this issue. It is reasonable to claim that the current standard of living in developing countries is much greater than it would have been without the industrialisation of developed countries. It is also reasonable to claim that the benefits of this process have largely accrued to those countries that undertook this process. This represents an empirical disagreement about how much different actors have benefitted from historical emissions. At some future time, it might be possible to determine exactly how each actor has benefited from historical emissions (for example through improved empirical analysis). But this information is not currently available and it is necessary to make a judgement about how much the benefits of emissions accrue to different agents. At the same time, there are different reasonable interpretations of this fact. Whilst there might be agreement about the fairness of Historical Responsibility principle, as well as the suitability of attributing responsibility to those who benefit from emissions, there are different reasonable positions regarding the extent to which actors benefit from spillover effects.

2.3.6 Equal Costs

Reasonable disagreement arises in another common allocation rule, the Equal Cost proposal. The Equal Cost proposal allocates emission rights so that each actor faces the same burden from limiting their emissions.[37] This requires allocating emission rights according to the financial resources that an agent has and, by framing

---

[35]For discussion, see: Nozick 1974, p. 93–4.

[36]Risse 2008, p. 30; Matthias Risse argues that there is also a diffusion of the benefits generated from historical emissions as developing countries benefit from technology created by previous generations of wealthy people.

[37]For further explication of this proposal, see: Ringius et al. 1998; Höhne et al. 2002, p. 52. One might contend that the Equal Cost proposal is Grandfathering, or Equal Welfare under a different guise. After all, each of these allocation rules ensures that actors experience equal welfare changes from participating in an emissions permit scheme. However there are important differences between these rules. The Equal Cost proposal allocates emission rights so that each agent faces the same cost of mitigation, because each should face an equal cost. This is different from Grandfathering, which ensures that individuals are not disadvantaged for choices that they

the problem in terms of costs, this approach represents a move away from the allocation of a common pool resource towards the allocation of the costs associated with the regulation of a resource.[38] Like the other proposals it too is subject to some reasonable disagreement. On the one hand it represents an appealing way of conceptualising the problem of distributing emission rights, because it frames the issue in terms of overall costs, rather than the benefits that arise from the distribution of a resource. We do not often think of distributing pollution rights as the distribution of a benefit, so it seems strange that we should adopt this approach for the distribution of emission rights. By equalising the cost of mitigation between different agents, the Equal Cost proposal recognises that we value emission rights because of the welfare that they bring, as well as taking a forward looking view of the costs that will arise from meeting emissions reduction commitments. On the other hand, different interpretations of this principle can be reasonably upheld, meaning that there is interpretative disagreement (type 2 disagreement) about this proposal.

This is because there are at least two issues that are unresolved in the application of Equal Costs proposal.[39] First, it is not clear whether the Equal Cost proposal advocates equalising a proportionate, or absolute reduction in the cost of achieving climate mitigation. Neither of these interpretations of the Equal Cost proposal can be rejected outright. On the one hand, equalising the absolute costs of mitigation seems appealing because it does exactly what the allocation rule sets out to achieve, namely: to equalise the cost of mitigation for everyone. By giving each actor the same absolute cost of mitigation, no single actor is privileged over another. On the other hand, equalising relative costs takes into account the relative differences that exist between actors. This may be important if greater levels of wealth (or any other metric that is being equalised) mean that some actors are able to cope with a certain cost better than others. Both of these proposals can be reasonably upheld for different reasons. One might reasonably argue that each actor should receive the same absolute cost regardless of his or her background conditions. Alternatively one might reasonably hold that the equal cost proposal should take into account background effects that impact the way we experience a cost. Neither of these proposals can be rejected outright.

Second, different claims can be made regarding what it is that should constitute a cost in this case. The Equal Cost proposal specifies the equalisation of burdens across actors, but it does not state how these burdens should be conceptualised. For instance, one might propose equalising economic costs, meaning that emission rights are distributed so that each actor experiences the same percentage reduction of economic wealth.[40] Alternatively, equalising welfare costs would allocate emissions

---

could not control. It is different from Equal Welfare because it allocates emission rights on the basis that mitigation is a cost, rather than a benefit.

[38]See: Meyer and Roser 2006, p. 229.

[39]Phylipsen et al. 1998; Traxler 2002.

[40]Burtraw and Toman 1993; Banuri et al. 1995, p. 105; Bodansky and Chou 2004, p. 30.

in a way that would equalise the costs of mitigation in terms of welfare.[41] It is not immediately apparent which of these proposals should be adopted. The equal economic cost proposal seems fair in one sense, but it disregards important elements of the differences in welfare between actors. Both of these positions can be reasonably upheld, and neither of these different ways of thinking about what a cost is can be reasonably rejected.

This implies that there are two areas of contention when distributing emission rights based on the Equal Cost proposal. On the one hand, it is not clear whether the Equal Costs proposal requires the equalisation of proportionate, or absolute costs. On the other hand, it is also unclear what a cost is in this case. In both cases, different views can be reasonably upheld, and there is likely to be disagreement about the application of this rule in practice. Because this represents disagreement about how the Equal Cost proposal should be interpreted in practice, this represents interpretative (type 2) disagreement.

2.3.7 Multi-criteria Approaches

Thus far, this chapter has discussed the relative fairness of different allocation rules in isolation from one another. But often these allocation rules are combined with one another in order to create Multi-criteria rules that distribute emissions using a combination of different allocation rules.[42] For example, the Greenhouse Development Rights Framework (GDR) distributes emissions rights according to an index, which is defined by a combination of the Historical Responsibility principle and a sufficientarian conception of justice (i.e. emissions should be allocated to enable the poor to develop).[43] Other Multi-criteria proposals include the Contraction and Convergence proposal,[44] which combines the proposals for Grandfathering and Equal per Capita Rights, and the Common but Differentiated Convergence proposal, which incorporates elements of the Equal Rights and Historical Responsibility proposals.

As a result, this chapter has drawn attention to reasonable disagreement about: the core principle that should be adopted for the fair distribution of emission rights, how this principle is interpreted, and the background facts that are important to the application of a certain allocation rule. But there is also reasonable disagreement about the relative importance that people should give to different allocation rules. Given that allocation rules are not typically considered in isolation, but rather as Multi-criteria rules, it is necessary to make a judgement about the importance

[41] See: Babiker and Eckhaus 2002.

[42] For a discussion of Multi-criteria approaches, see: Metz et al. 2002.

[43] Baer et al. 2009; Kanie et al. 2010, p. 306.

[44] See: Meyer 2007.

that each rule should receive. There are different reasonable positions about the weighting each allocation rule should be given. It is clear that it is unreasonable to prioritise some rules above others. For instance, it is unacceptable to argue that Historical Responsibility should take precedent over the Basic Emission Rights proposal, because allowing people to meet their basic human needs is far more pressing than correctly attributing responsibility. But in other cases the choice is much less clear-cut. For instance, whilst it's reasonable to propose that people should put a high importance on the Historical Responsibility principle, there's no reason why someone can't reasonably contend that people should give equal importance to the Equal Rights proposal. There are different reasonable positions about the significance that each allocation rule should receive, none of which can be reasonably rejected. As a result, specifying a Multi-criteria approach does not escape the problem of reasonable disagreement.

2.3.8 Reasonable Disagreement Over the Fair Distribution of Emissions

Following the discussion of disagreement given in Sect. 2.2, this section has shown the following. Foremost, there are different reasonable claims about which allocation rule should be used for the fair distribution of emission rights. This is normative disagreement about the core principle that should be used to distribute emission rights fairly. Second, even if there is agreement on which core principle should be applied in this context, there are different reasonable claims about the way that each principle is applied. On the one hand, applying or interpreting each allocation rule requires some judgement about how this allocation rule should be applied (interpretative disagreement). On the other hand, certain elements of each allocation rule depend on facts that are not specified by the allocation rule itself (normative disagreement and epistemic disagreement). Third, there are different reasonable claims about the relative weight that each principle should be given (weighting disagreement).[45] This account of disagreement is not exhaustive, and there may be other types of disagreement here. Most notably, I've not discussed methodological disagreement surrounding climate change, which may lead to even greater problems when trying to reach consensus in the UNFCCC. But the account that I've given here shows that there are different reasonable claims regarding the fair distribution of emission rights, even when there is agreement about other central issues, including agreement over the correct methodology to apply. Following the discussion in Sect. 2.2, this means that actors cannot dismiss these views as unreasonable, even if they deny that these views are correct. The upshot of this is

---

[45]One issue that I've not considered here is methodological disagreement. Whilst methodological disagreement is very relevant, this section has shown that is that reasonable disagreement can exist even if there is agreement over the methodology that is employed.

that the fair distribution of greenhouse gas emission rights is subject to considerable reasonable disagreement. As I show in Chap. 3, this has significant implications for the design of the decision-making procedures of the UNFCCC.

2.4 Reasonable Disagreement in Theory and Practice

I've shown that there are different reasonable claims about the fair distribution of greenhouse gas emissions. I've not stated what the relevance of this disagreement is, nor have I shown that there is reasonable disagreement in the UNFCCC itself. In the following chapter, I go on to consider how actors should respond to reasonable disagreement. But before doing this it is necessary to state how pervasive reasonable disagreement actually is in the UNFCCC.

Following the discussion in Sect. 2.2, the fact that different proposals for the fair distribution of emission rights are reasonable means that none of these can be dismissed as unreasonable interpretations of what a fair distribution is. Whilst an actor can reject a proposal as an incorrect account of what a fair distribution is, they cannot reject a reasonable proposal as something that doesn't deserve our respect. One might acknowledge that a claim is reasonable, whilst disagreeing that it is a correct answer to the problem at hand. But by accepting that a claim is reasonable, an actor recognises that the claim should be treated with adequate respect, even if the actor disagrees with the claim itself. Because people can reject different reasonable proposals as correct accounts of what is fair, it isn't possible to say with any certainty whether any of these proposals will gain the support of all reasonable actors. In theory, this means that deliberation is unlikely to bring about agreement if there are many heterogeneous actors. Different actors can support different reasonable positions, and it isn't possible to reject these as unreasonable interpretations of what is correct. This implies that no single proposal will gain the support of reasonable persons.

This is important because identifying a fair distribution of emission rights is a necessary element of many policies to mitigate greenhouse gas emissions. Whilst I give a full account of this argument in Chap. 3, it's worth saying something about it here. Climate change is caused by actions that infiltrate almost every element of society on a global scale. Further, the causes and consequences of these actions have profound implications for people's welfare. Policies to reduce emissions are therefore likely to bring about very large redistributions of welfare. Generally speaking, people are only prepared to endorse policies that they perceive as fair. Further, the fair distribution of emission rights reflects concerns about fair burden sharing that are pervasive throughout climate change policy more generally. What this means is that reasonable disagreement about the fair distribution of emission rights precludes us from adopting mitigation policies, even when there is agreement on other issues, such as the need to limit global temperature rises to no more than 2 °C above pre-industrial levels, or on the best mitigation policy for achieving this end.

2.5 Reasonable Disagreement in the UNFCCC

Whilst this means that there is reasonable disagreement in theory, it is also necessary to give an account of whether this disagreement exists in practice. The first thing to say is that many of the principles and claims mentioned in this section are proposed (and rejected) in climate change negotiations for the fair distribution of emission rights. For instance, the following is a brief account of the state delegations that have proposed specific allocation rules in the history of the UNFCCC negotiations:

1. Equal per capita: France, Switzerland, India[46]
2. Equal costs: Poland, Australia[47]
3. Historical Responsibility: Brazil[48]
4. Multi-criteria: UK, Germany[49]

This is just a limited account of how some of these allocation rules are advocated in practice. The important point is that this is not just a matter of philosophical discussion, but also a problem in practice. One might take a cynical view of state behaviour in international negotiations and argue that these actors adopt positions that best represent their own self-interest. Given that it isn't possible to know why a state delegation makes a proposal, it is not possible to refute this claim. However, people from academic and policy circles have been thinking about and discussing this issue for over two decades. Unlike state delegates, many of these authors have an impartial view about which countries should benefit from a system of emission permits, yet there is still significant and persistent disagreement about which principle to adopt.

Before concluding this section, it is worth briefly saying something about why people disagree about the fair distribution of emission rights. Following the earlier discussion in this chapter, there is disagreement over the fair distribution of emission rights because different world views lead us to form different judgements about which allocation rule should be adopted. These different experiences and world views also lead to disagreement about how these principles should be applied, how people should judge empirical evidence relevant to the allocation rule, and the relative weight that people should give these principles. Each of these issues requires a judgement on the part of the decision-maker. Given that the ability to judge and interpret is different for each actor, different positions are likely to arise even if each actor is acting reasonably. It is the fact that disagreement arises out of different interpretations of what is fair, rather than from mistake or self-interest, which requires us to respect each reasonable view.

---

[46] Agarwal and Narain 1991; AGBM 1996.

[47] Müller 1999, p. 6 (note that Benito Müller refers to these as per capita GDP and per capita economic welfare).

[48] AGBM 1996; Brazilian Party 1997.

[49] Garnaut Report 2009, p. 203; Ross Garnaut notes that both the UK and Germany advocate the Contraction and Convergence rule, which is a Multi-criteria rule.

2.6 Summary

This chapter has shown that there are different reasonable positions regarding the fair distribution of emission rights, and that there is disagreement on this issue in practice. This is not to say that no independent criterion that commands the support of all reasonable parties for the fair distribution of emission rights exists. Rather, no such criterion is immediately identifiable for specifying what a fair distribution of emission rights is. It might be the case that sufficient discussion and debate about substantive fairness may eventually lead parties to reach agreement on these issues. But it seems unlikely that actors in climate change institutions will reach agreement on the fair distribution of emission rights at any point in the near future. Given the need to achieve action on climate change quickly, it seems reasonable to assume that either no independent criterion for substantive justice exists or, if it does exist, that it is (for all purposes) unknown and indeterminate.

References

Agarwal, A., and S. Narain. 1991. *Global warming in an unequal world, a case of environmental colonialism*. New Delhi: Centre for Science and Environment.

AGBM. 1996. *Quantified emission limitation and reduction objectives within specified time-frames: Review of possible indicators to define criteria for differentiation among annex I parties*. FCCC/AGBM/1996/7.

Allen, M.R. 2009. Planetary boundaries: Tangible targets are critical. *Nature Reports Climate Change* 3(0910): 114–115.

Allen, M.R., D.J. Frame, et al. 2009a. Commentary: The exit strategy. *Nature Reports Climate Change* 3: 56–58.

Allen, M.R., D.J. Frame, et al. 2009b. Warming caused by cumulative carbon emissions towards the trillionth tonne. *Nature* 458: 1163–1166.

Arrhenius, S. 1896. On the influence of the carbonic acid in the air upon the temperature of the ground. *Philosophical Magazine and Journal of Science* 41(251): 237–276.

Babiker, M. H., and R.S. Eckhaus. 2002. Rethinking the Kyoto targets. *Climatic Change* 54: 99–114.

Baer, P., and T. Athanasiou. 2007. *Frameworks & proposals; a brief, adequacy and equilty-based evaluation of some prominent climate policy framworks and proposals*. Heinrich Boll Foundation Global Issue Papers, No. 30. Heinrich Boll Foundation.

Baer, P., G. Fieldman, et al. 2009. Greenhouse development rights: A proposal for a fair global climate treaty. *Ethics, Place & Environment* 12(3): 267–281.

Banuri, T., K. Goran-Maler, et al. 1995. Equity and social considerations. In *Economic and Social Dimensions of Climate Change. Contribution of Working Group III to the Second Assessment Report of the Intergovernmental Panel on Climate Change*, ed. J.P. Bruce, L. Hoesung and E. Haites. Cambridge: Cambridge University Press.

Berk, M.M., and M.G.J. den Elzen. 1998. The Brazilian protocol evaluated. *Climate Change* 44: 19–23.

Besson, S. 2005. *The Morality of Conflict: Reasonable Disagreement and the Law*. Oxford: Hart Publishing.

Bodansky, D., and S. Chou. 2004. *International climate efforts beyond 2012: A survey of approaches*. Washington, DC: Pew Center on Global Climate Change.

Bovens, L. 2011. A lockean defense of grandfathering emission rights. In *The ethics of global climate change*, ed. D.G. Arnold. Cambridge: Cambridge University Press.

Brazilian Party. 1997. *Proposed elements of a protocol to the United Nations Framework Convention on Climate Change*. Presented by Brazil in response to the Berlin Mandate at the United Nations Framework Convention on Climate Change.

Burtraw, D., and M.A. Toman. 1993. Equity and international agreements for CO2 constraint. *Journal of Energy Engineering* 118(2): 122–135.

Caney, S. 2005. Cosmopolitan justice, responsibility and global climate change. *Leiden Journal of International Law* 18: 747–745.

Caney, S. 2006. Environmental degradation, reparations, and the moral significance of history. *Journal of Social Philosophy* 73(3): 464–482.

Caney, S. 2009. Justice and the distribution of greenhouse gas emissions. *Journal of Global Ethics* 5(2): 125–146.

Caney, S. 2012. Just emissions. *Philosophy and Public Affairs* 40(4): 255–300.

den Elzen, M. 2002. *Exploring post-Kyoto climate regimes for differentiation of commitments to stabilize greenhouse gas concentrations*. Climate Change Policy Support Project, Dutch Ministry of Environment.

Dworkin, R. 1981. What is equality? Part II equality of resources. *Philosophy and Public Affairs* 10(4): 283–345.

Fuji, Y. 1990. An assessment of the responsibility for the increase in CO2 concentration and inter-generational carbon accounts. *International Institute for Applied Systems*. Working Paper 90–55.

Gardiner, S. 2004. Ethics and global climate change. *Ethics* 114: 555–600.

Garnaut, R. 2009. *The Garnaut report*. Great Britain: Economic and Social Research Council.

Gosseries, A. 2003. Historical emissions and free-riding. In *Historical justice*, ed. L. Meyer. Nomos: Baden-Baden.

Gosseries, A. 2005. Cosmopolitan luck egalitarianism and climate change. *Canadian Journal of Philosophy Supplementary* 31: 279–309.

Griffin, J.W. 2008. *On human rights*. Oxford/New York: Oxford University Press.

Grubb, M. 1992. Options for an international agreement. In *Combatting global warming: Study on a global system of tradeable carbon emission entitlements*, ed. M. Grubb and A. Rose. Geneva: UNCTAD.

Grubb, M. 1995. Seeking fair weather: Ethics and the international debate on climate change. *International Affairs* 71(3): 463–496.

Gupta, S., D.A. Tirpak, et al. 2007. Policies, Instruments and Co-operative Arrangements. In *Climate change 2007: Mitigation. Contribution of working group III to the fourth assessment report of the Intergovernmental Panel on Climate Change*, ed. B. Metz, O.R. Davidson, P.R. Bosch, R. Dave, and L.A. Meyer. Cambridge: Cambridge University Press.

Gutmann, A., and D. Thompson. 1996. *Democracy and disagreement*. Cambridge, MA: Harvard University Press.

Hart, H.L.A., and T. Honor. 1985. *Causation in the law*. Oxford: Clarendon.

Heyward, M. 2007. Equity and international climate change negotiation: A matter of perspective. *Climate Policy* 7: 518–534.

Höhne, N., C. Galleguillos, et al. 2002. *Evolution of commitments under the UNFCCC: Involving newly industrialized economies and developing countries*. The German Federal Environmental Agency (Umweltbundesamt).

Jacoby, H.D., et al. 1999. *Toward a useful architecture for climate change negotiations*. Cambridge, MA: MIT Joint Program on the Science and Policy of Global Change.

Kanie, N., H. Nishimoto, et al. 2010. Allocation and architecture in climate governance beyond Kyoto: Lessons from interdisciplinary research on target setting. *International Environmental Agreements: Politics, Law and Economics* 10(4): 299–315.

La Rovere, E.L., L. Valente de Macedo, et al. 2002. The Brazilian proposal on relative responsibility for global warming. In *Building on the Kyoto Protocol: Options for protecting the climate*, ed. K. Baumert, O.C. Blanchard, S. Llosa and J.F. Perkaus. Washington: World Resource Institute.

Mackie, J.L. 1980. *The cement of the universe: A study of causation*. Oxford: Oxford University Press.

McMahon, C. 2009. *Resonable disagreement*. Cambridge: Cambridge University Press.

Meinshausen, M., N. Meinshausen, et al. 2009. Greenhouse-gas emission targets for limiting global warming to 2 °C. *Nature* 458(7242): 1158.

Metz, B., M. Berk, et al. 2002. Towards an equitable global climate change regime: Compatibility with Article 2 of the climate change convention and the link with sustainable development. *Climate Policy* 2(2–3): 211–230.

Meyer, A. 2007. The case for contraction and convergence. In *Surviving climate change*, ed. M. Levene and D. Cromwell. London: Pluto Press.

Meyer, L., and D. Roser. 2006. Distributive justice and climate change. The allocation of emission rights. *Analyse & Kritik* 28: 223–249.

Miller, D. 2008. Global justice and climate change: How shall responsibilities be distributed? In *Tanner lectures on human values*. Delivered at Tsinghua University, Beijing, March 24–25, 2008.

Müller, B. 1999. Justice in global warming negotiations: How to obtain a procedurally fair compromise *Oxford Institute for Energy Studies*. EV26.

Neumayer, E. 2000. In defence of historical accountability for greenhouse gas emissions. *Ecological Economics* 33: 185–192.

Nozick, R. 1974. *Anarchy, State and Utopia*. New York: Basic Books.

Pan, J. 2003. Emissions rights and their transferability. Equity concerns over climate change mitigation. *International Environment Agreements: Politics, Law and Economics* 3(1): 1–16.

Phylipsen, G.J.M., et al. 1998. A triptych sectoral approach to burden differentiation; GHG emissions in the European bubble. *Energy Policy* 26(12).

Pogge, T.W. 2005. World poverty and human rights. *Ethics & International Affairs* 19(1): 1–7.

Rawls, J. 1993. *Political liberalism*. New York: Columbia University Press.

Rayner, S., E. Malone, et al. 1999. Equity issues and integrated assessment. In *Fair weather? Equity concerns in climate change*, ed. F. Toth. London: Earthscan.

Ringius, L., et al. 1998. Can multi-criteria rules fairly distribute climate burdens – OECD results from three burden sharing rules. *Energy Policy* 26(10): 777–93.

Ringius, L., et al. 2001. Burden sharing and fairness principles in international climate policy. *International Environmental Agreements: Politics, Law and Economics* 2: 1–22.

Risse, M. 2008. Who should shoulder the burden? Global climate change and common ownership of the earth. In *Harvard Kennedy School Faculty Research Working Paper Series*. No. RWP08-075.

Scanlon, T. 1998. *What we owe to each other*. Cambridge/London: Belknap Press of Harvard University Press.

Shue, H. 1995. *Avoidable necessity: Global warming, international fairness, and alternative energy*, Theory and practice: NOMOS XXXVII. New York/London: New York University Press.

Shue, H. 1996. Basic rights: Subsistence, affluence, and U.S. foreign policy. Princeton University Press.

Shue, H. 1999. Global environment and international inequality. *International Affairs* 75(3): 531–545.

Singer, P. 2002. *One world: The ethics of globalization*. New Haven/London: Yale Nota Bene.

Solomon, B.D., and D.R. Ahuja. 1991. International reductions of greenhouse gas emissions; an equitable and efficient approach. *Global Environmental Change* 1(5): 343–350.

Traxler, M. 2002. Fair chore division for climate change. *Social Theory and Practice* 28: 101–34.

Waldron, J. 1999. *Law and disagreement*. Oxford/New York: Oxford University Press.

Chapter 3
Reaching Agreement Through Fair Process

3.1 Introduction

Chapter 2 showed that there is reasonable disagreement over some of the ends that the UNFCCC should achieve. This chapter builds on this finding by showing that, where there is reasonable disagreement over such ends, fair decision-making processes gain additional importance. That is, this chapter makes a case for the importance of fair procedures in the UNFCCC. It does so in three steps. First, it discusses the relative merits of fair procedures, arguing that, whilst fair procedures are important in themselves, there are sometimes trade-offs between designing a process that is procedurally fair and designing a process to meet other more pressing ends. Second, it argues that, whilst there is disagreement in the UNFCCC, there is also agreement over some of its ends, most importantly, that it should collectively limit emissions to avoid causing dangerous climate change. Further, given certain specific characteristics associated with climate change, achieving this goal requires meeting certain criteria, including stringency, urgency, and voluntary cooperation. This means that it is important to find a way of reaching agreement in the UNFCCC even when there is reasonable disagreement over some of the ends that it should bring about. Third, given these specific characteristics and requirements, and given the existence of reasonable disagreement, I show that fair procedures theoretically provide a way of reaching agreement in the UNFCCC even when there is reasonable disagreement over the ends that it should pursue.

3.2 The Importance of Fair Procedures

There are three reasons for designing decision-making processes in ways that honour procedural values. One of these reasons appeals to the intrinsic nature of procedural values, whilst the other two appeal to their instrumental value.

L. Tomlinson, *Procedural Justice in the United Nations Framework Convention on Climate Change*, DOI 10.1007/978-3-319-17184-5_3

First, procedural values are important in themselves, regardless of the outcomes that a procedure brings about.[1] Whilst one might consider the outcome of a certain decision-making process to be fair, people often have serious reservations if an outcome comes about in an unfair way. Empirical studies show that this is true for human behaviour generally, as well as in the UNFCCC itself. In two often cited sources, Thibaut and Walker, and Walker et al., use comparative empirical analysis to show that disputants in legal proceedings are often as concerned about the fairness of the process by which an outcome is reached as they are about the fairness of the outcome itself (Thibaut and Walker 1975; Walker et al. 1979). In the UNFCCC, state delegates often make demands for procedural fairness and democratic participation.[2] For example, some authors suggest that an absence of procedural values played a part in the decision of several states to reject the Copenhagen Accord in 2009 (Eckersley 2012, p. 33; Bodansky and Rajamani 2013, p. 12). At these negotiations, Venezuela, Cuba, Nicaragua and Bolivia all renounced the Copenhagen Agreement on procedural grounds.[3] At the COP16 negotiations, representatives of Yemen, acting on behalf of the Group of 77 and China, called for more transparent and inclusive decisions.[4] There are many concerns about the inequality of resources between delegations in climate negotiations, including financial resources, technical expertise and scientific information.[5] Others have questioned the legitimacy of the G8 and MEF on procedural grounds, arguing that they exclude key actors (Karlsson-Vinkhuyzen and McGee 2013, p. 67). It is clear that actors are not content to accept outcomes based on processes that are unfair, and that procedural values are important here.

Against this view, one might argue that the relative importance of procedural values in the UNFCCC remains disputable. Some authors argue that there is no intrinsic merit to procedural design and that decision-making processes should be designed with the sole intention of promoting certain desirable outcomes. For instance, Richard Arneson argues that democracy should be regarded as ‘a tool or instrument that is to be valued not for its own sake but entirely for what results from having it’ (Arneson 2004). According to these arguments, decision-making processes should be designed to achieve desired outcomes, rather than to promote values that are independent of these ends. Even if one accepts that procedural values are important, it might be the case that they should be trumped by other, more pressing concerns. Procedural values often conflict with substantive ones. In multilateral institutions, for example, concerns for fair procedure often lead to large numbers of actors participating in decisions and discussions, which can hold

[1]Several theorists hold that procedures are intrinsically important: Beitz 1989; Christiano 1996; Waldron 1999.

[2]See: Lange et al. 2007; Bäckstrand 2010, p. 1; Eckersley 2012.

[3]For discussion, see: Dubash 2009, p. 8; Bäckstrand 2010, p. 1.

[4]See: IISD 2010, p. 3.

[5]For example: Agarwal and Narain 1991; Bruce et al. 1995, p. 84; Gupta 2000; Chasek and Rajamani 2003.

up process and lead to conflict and disagreement. Given the high stakes involved in climate change, one might ask why the UNFCCC should concern itself with procedural values, if there are more pressing substantive ends that it could pursue. Whilst the intrinsic nature of procedural values provides some justification for considering these values in the UNFCCC, further explanation is needed as to why people shouldn't prioritise important substantive ends. I show this by appealing to two further reasons for pursuing procedural values, which are based on their instrumental value towards achieving other ends.

The second reason that procedural values are important rests on two points: (i) there is an urgent need to implement collective action on climate change, and (ii) cooperative action on climate change is dependent on consensual agreement among a sufficient number of actors. Points (i) and (ii) rest on empirical assumptions that I discuss fully Sect. 3.3. In order to avoid getting bogged down in lengthy arguments, for the moment I simply assume that these are relatively uncontroversial assumptions. If there is a need to implement action on climate change quickly, and a need for consensual agreement among actors, then this gives a further reason for thinking that fair procedures are important. Actors are generally reluctant to participate in agreements that they perceive to be unfair, either to themselves or to other actors. If there is a pressing need to reach agreement on climate change quickly, then fair procedures are important for ensuring the cooperation of all relevant actors.

But this reason for valuing fair procedures does not show why fair procedures are important when there are other more pressing ends. In the same way that one might prioritise more pressing ends above intrinsic value of procedural fairness, one might also think that more pressing ends should take priority over this instrumental concern for fair process. Designing decision-making procedures that are fair can sometimes lead to trade-offs with other values. If, for example, fair procedures are those in which a large number of actors can participate, then this may prevent actors from reaching a meaningful agreement quickly, thereby conflicting with other important ends. Climate change is a problem that is sometimes thought to have potentially bad, or even catastrophic outcomes.[6] In light of these sorts of concerns, it might make more sense to design decisions so that they bring about a particular important end, rather than to worry about procedural concerns. Even if actors sometimes consider procedural fairness to be important, these concerns may seem less relevant if we can state what a fair outcome is and how to reach that outcome. One might easily think that where procedural values clash with other more important ones, the UNFCCC should forego concerns for procedural fairness for the sake of avoiding very bad outcomes. This means that the first two reasons that I've given for valuing fair procedures are insufficient for showing why procedural values are important elements of the UNFCCC, given that climate change can bring about such negative outcomes.

[6]For more on the potentially catastrophic nature of climate change, see: Schneider and Lane 2006.

But there is a third reason to think that procedural values are important. Consider someone who asks why these processes should be designed to meet procedural values, rather than to achieve fair substantive ends. If we know what a fair outcome is then we can specify that the procedure should achieve this end, rather than concern ourselves with procedural values that may conflict with other important matters. Our third reason gives an answer to this point. Procedural values are important for the design of the UNFCCC because it is not possible to specify what substantive values a procedure should adopt when there is reasonable disagreement about these substantive values. In addition to points (i) and (ii), this third reason requires: (iii) that there is reasonable disagreement over the substantive ends that the institution should achieve and, (iv) that it isn't possible to reach consensus among actors who reasonably disagree by focussing on substantive ends. I've already defended points (iii) and (iv) in Chap. 2. Points (i) and (ii) are the subject matter of Sects. 3.3 and 3.3.1, respectively. Taken together, these points suggest that procedural values are important in the UNFCCC. If people can't specify the outcomes that the UNFCCC should achieve, then it may be possible to achieve an outcome that all can agree is fair by using a fair procedure.

3.3 Avoiding Dangerous Climate Change

Whilst Chap. 2 showed that there is some disagreement about the correct ends that the UNFCCC should pursue, this doesn't mean that there's disagreement over all of its ends. In fact, the convention text of the UNFCCC outlines the collective aims that its participants have collectively agreed to. For example, the guiding principle of the UNFCCC is to stabilise atmospheric concentrations of greenhouse gases at a level that avoids dangerous climate change (UNFCCC Article 2). Whilst there are still different ideas about what this means in the academic literature, there is now some agreement in the UNFCCC that avoiding dangerous climate change means limiting global temperature changes from climate change to no more than 2 °C above pre-industrial levels.[7] This goal is affirmed in the Copenhagen Accord, the Cancún Agreements, the Durban Platform, and the Lima Call for Climate Action as well as in other multilateral agreements such as the Declaration of the Leaders of the MEF.[8] The 2 °C target is therefore something that receives a lot of support within the UNFCCC itself and the implications of exceeding this target are so severe that missing it means endangering many other policy goals.

Chapter 2 argued that, where there is reasonable disagreement, parties are unlikely to be able to reach agreement despite continued deliberation. The rest of this section argues that this is problematic because avoiding dangerous climate

[7]Not everyone agrees with this view, and some argue that a more stringent temperature target is needed in order to avoid dangerous climate change (see: Heyward 2007; Steinacher et al. 2013).

[8]See: UNFCCC 2009, 2010, 2011, 2014; MEF 2009.

change requires a global cooperative effort that meets certain important criteria. That is, avoiding dangerous climate change requires action that is urgent, and under conditions of voluntary consensual agreement (points (i) and (ii) from earlier). Having done this, I then go on to show that the combination of these factors means that fair procedures are a necessary feature of the UNFCCC if it is bring about sufficient action to avoid dangerous climate change.

3.3.1 Stringency

Before showing that climate change is a matter of urgency, it's necessary to say something about what is required to meet the 2 °C target. Given that there is a general understanding that UNFCCC should limit global temperature changes to no more than 2 °C above pre-industrial levels, it's now possible to consider what this actually requires in terms of collective action. Climate change is caused by actions that produce greenhouse gas emissions through the burning of fossil fuels or changes in land use such as deforestation. Leaving aside the possibility of geoengineering, which is still unproven, maintaining global temperature changes within the 2 °C threshold will require reductions of emissions on a global scale. The question now becomes what level of emissions reduction is needed to meet the 2 °C target.

This question is addressed in two articles published in the journal *Nature* in 2009. In one article Meinshausen et al. argue that people should look at cumulative emissions to the year 2050 as an indicator of whether the world will stay within the 2 °C limit by 2100 (Meinshausen et al. 2009). Taking this approach, the authors provide emissions targets for achieving specific temperature increases by 2100. In a second article, Allen et al. argue that the impact of emissions should be considered over a much longer time period and provide emissions targets for limiting atmospheric warming until 2500 (Allen et al. 2009b). Both of these articles suggest that meeting the 2 °C limit requires limiting our total cumulative emissions from the industrial revolution to less than a trillion tonnes of carbon dioxide. Both studies suggest that meeting the target requires large cuts in global emissions. This means that there is a 'carbon budget' that cannot be exceeded if people are serious about remaining within the 2 °C temperature threshold. Other research supports the findings of these studies. For example, the IPCC also suggests that meeting this temperature target requires that global emissions be reduced by 50–85 % by 2050 relative to 1990 levels (Gupta et al. 2007, p. 775).

This gives one requirement of an effective climate agreement: stringency. Stringency implies that global emissions of greenhouse gases are reduced in accordance with the pathways defined by Meinshausen et al., Allen et al., and the IPCC. Limiting global average temperature increase to no more than 2 °C means enforcing stringent emissions reductions.

The problem is that the world is not on track towards meeting this end. Whilst the international community has made a number of pledges to reduce emissions, these are inadequate for what is required. In 2011, a report from the International Energy

Agency (IEA) states that current emissions trends are consistent with a long-term temperature increase of more than 3.5°carbon (IEA 2011).[9] As it is, the 2 °C target required for climate stabilisation is unlikely to be met.

3.3.2 *Urgency*

But avoiding dangerous climate change is as much about the rate at which emissions are reduced as it is about the stringency with which this is done. There are several reasons why this is the case.[10]

Firstly, a lot has already been emitted. Since the industrial revolution, the world has emitted a cumulative total of around half a trillion tonnes of carbon dioxide (Allen et al. 2009a). This has pushed up global atmospheric concentrations of greenhouse gases to 400 ppm, which is the highest that they have been in at least the past three million years. The fact that so much has already been emitted has two important implications. One related to adaptation and one to mitigation.

On the one hand, climate change is already happening and will continue to get worse in the future.[11] It is well documented that the earth's atmosphere is undergoing extreme changes and that the possibility of limiting these changes diminishes as the starting point for action is pushed further into the future (Gardiner 2011; Stocker 2013). The greenhouse gases that have been put into the atmosphere since the industrial revolution have already caused an increase in the global average surface temperature of around 1 °C (IPCC 2013). This has caused changes to the climate that are having severe impacts on human welfare globally.[12] This problem is exacerbated by the fact that the physical effects of climate change are time-lagged. Due to inertia within the atmospheric system, the full effects of emissions that are generated today are not translated into physical changes in the atmosphere until sometime in the future. This also means that those emissions that were generated in the past will continue to cause an increase in global temperatures regardless of the actions that people take from now on.[13] Given that climate change is already happening and will continue to happen in the future, any attempt to mitigate the very worst effects of climate change must take place as soon as possible.

On the other hand, there is a further issue associated with the fact that the world has already emitted a lot, which is that the atmospheric concentration of greenhouse gases is already close to the threshold limit for avoiding dangerous

[9]See also, den Elzen 2010.

[10]These points draw from those in Allen et al. 2009a; Stern 2014.

[11]For discussion, see: Hare and Meinshausen 2006; Allen et al. 2009a.

[12]See: Patz et al. 2005; Human Rights Council 2008.

[13]See: Hare and Meinshausen 2006; Rogelj et al. 2011.

climate change.[14] Given the intensity with which people are currently burning fossil fuels, it is expected that this threshold will be reached within the next two decades (Oliver et al. 2012). This means that the opportunity to prevent severe climate change is rapidly diminishing. If people are serious about staying within the 2 °C limit then action on climate change needs to happen soon.

A second reason for urgency is that the stakes are high and the outcomes unknown. Climate science is characterised by high uncertainty. The atmosphere is extremely complex and the implications of changes in atmospheric concentrations of greenhouse gases are difficult to model. Despite our best efforts to make predictions about climate change there is still a great deal of uncertainty about what its full effects will be in the future. This is particularly important given that climate change may bring about consequences that are non-linear, irreversible, and potentially catastrophic in nature.[15] The effects of emissions are non-linear because there are both tipping points and positive feedback loops within the earth's climate. This means that any given amount of emissions may lead to far greater changes in the atmosphere than the same amount of emissions brought about in the past. Some of these tipping points may bring about changes that are wholly irreversible (Halsnæs et al. 2007, p. 127; Solomon et al. 2009; Frolicher and Joos 2010). Climate change is also potentially catastrophic. The atmospheric changes that come about from unabated emissions could lead to a situation in which larger parts of the planet become uninhabitable. Even if the likelihood of such outcomes arising is very small, there are still strong arguments for reducing emissions as quickly as possible (Weitzman 2009). When these three features of climate change are combined, it is difficult to argue against taking immediate action to mitigate global emissions.

A third reason is that the ability to implement necessary measures to avoid dangerous climate change diminishes with time. This is for at least two separate reasons. On the one hand, there is technical inertia in implementing emissions reductions. The total amount of emissions that can be reduced in a certain amount of time is finite (den Elzen et al. 2006, p. 7). Carbon emissions are fundamental to every aspect of global society and making the necessary changes to global energy infrastructure to bring about a reduction in emissions will take a long time. On the other hand, 'political' or 'social' inertia is also an issue here.[16] Energy infrastructure operates on very long time-scales, meaning that any decisions that are made now are likely to have implications that last for a very long time. Power plants, fuel storage facilities, and electricity grids have operational lifetimes of several decades. As a result, some authors argue that decisions on energy policy run the risk of 'lock-in', whereby decisions taken today have long-term implications for carbon emissions (Unruh 2000; IEA 2011, p. 2). This problem is worsened by the large

[14]Several authors argue that argue that the 2 °C target will soon become unachievable: IPCC 2007; Peters et al. 2013, p. 5.

[15]Schneider and Lane 2006; For climate change tipping points see: Lenton 2011; Voorhar and Myllyvirta 2012; for abrupt climate change, see: Alley et al. 2005.

[16]See: Matthews and Solomon 2013.

costs involved with the development of energy infrastructure. Any decisions that are made now will continue to have implications for the future because reversing these decisions is prohibitively costly. Achieving the large reductions necessary for climate stabilisation therefore requires action now because delaying action may lead us in a direction that can't be changed.

A fourth reason for urgency is that postponing action now makes future emissions reductions very costly.[17] This is partly due to the issues raised above. If a large proportion of the total carbon budget needed to stay within the 2 °C limit are 'locked-in' by existing energy infrastructure, then any additional infrastructure has to be entirely carbon free. This is likely to be much more costly than a mitigation policy that takes gradual steps towards reducing emissions, phasing out emissions intensive infrastructure over time rather than stopping all emissions abruptly.

But further to this, if action isn't taken to reduce the growth of global emissions now then even more stringent measures have to be adopted later on. It's not just the case that emissions aren't being reduced at the moment; global emissions are actually increasing and are expected to continue to do so in the future. The IEA predicts that global energy demand is likely to grow by over a third between now and 2035. Most of this growth is expected to come from increased energy consumption in developing countries that are undergoing rapid economic development. As a result of this increase in energy demand, energy-related carbon emissions are predicted to rise from an estimated 31.2–37.0 Gt in the same period (IEA 2011). Postponing action on mitigation until some point in the future means that the rate at which emissions have to be reduced will be higher than if this reduction starts now. A number of recent studies have made a strong economic case in support of more immediate action on climate change now in order to avoid more costly action later (Stern 2014; Global Commission on the Economy and Climate 2014). Given that the cost of reducing emissions increases in line with the amount that has already been emitted, postponing mitigation until some point in the future will ultimately involve higher costs.

This argument is based on the idea that it is economically efficient to take action on mitigation now rather than later. But some authors dispute this view, arguing that, rather than taking costly action to immediately reduce our emissions now, it is more cost effective to invest in clean technology for the future and postpone mitigation until suitable alternatives for producing energy are developed (Lomborg 2001). This is because what matters for staying below a certain global temperature target is the total cumulative amount of emissions up to a certain point in time, rather than the rate at which they are reduced.[18] Furthermore, the cost of mitigating emissions in the future can be discounted because technological developments and economic growth will means that those who are alive in the future are expected, on aggregate, to be better off than those who are alive today. Under this approach, the 2 °C target is still feasible if global emissions continue at their current levels, or even increase in the

[17]For discussions of this point, see: den Elzen et al. 2006; IEA 2011; Dirix et al. 2013.

[18]Allen et al. make this point, although not in support of Lomborg's argument (Allen et al. 2009a).

near future. Given that it is possible to develop clean ways of generating energy, the international community should invest in these technologies and reduce emissions in the future rather than taking costly mitigation measures now.

But this represents an extremely optimistic view of how things might play out. Whilst it is impossible to rule out the possibility of developing sufficient alternatives to emissions intensive energy production in the future, there are important reasons for thinking that this is unlikely. The world is already very close to exceeding the total amount of emissions needed to meet the 2 °C target. Given the physical inertia of the climate system and the problem of carbon lock-in, many think that global emissions need to peak in the immediate future if dangerous climate change is to be avoided. There simply isn't enough time to develop clean technologies whilst avoiding mitigation.

Further to this, the argument for postponing mitigation is based on the assumption that it is possible to develop cheap and reliable ways of producing clean energy. Yet, whilst there has been some investment in clean energy in the past few years, this is yet to produce a way of generating low carbon energy that is cost comparable with fossil fuels.[19] Clean energy technology is still extremely costly and suffers from technical problems such as intermittency. There is no guarantee that investment in clean energy will bring about the necessary advances in technology to maintain the current standard of living enjoyed by many people around the world. This doesn't mean that investment in research and development for clean energy should stop. Developing clean energy may ultimately be crucial for helping people live decent lives without causing dangerous climate change. But it may take a long time for the price of clean energy to fall and it shouldn't be taken for granted that these technologies will be developed in the near future. Continuing to produce emissions now in the hope that suitable alternative modes of energy production will come about in the future is therefore naïve. In fact, starting emissions reductions now rather than later delays the point at which certain temperature limits are reached.[20] This provides more time for developing clean energy technologies and building the necessary infrastructure for meeting energy needs in a low carbon world. If people are serious about developing clean technologies they should pursue mitigation policies now rather than later.

A third point that supports these two arguments is that the necessary technological advances in energy production may only come about once sufficient incentives are in place to develop them. One of the main problems with clean energy at the moment is that it is so expensive in comparison to fossil fuels. This makes it difficult to encourage people to develop and use these forms of energy production. One way of changing this situation is to provide sufficient incentives for people to move away from carbon intensive methods of energy production and towards low-carbon options. Mitigation policies can play an important role here. These policies increase the cost of using fossil fuels thereby encouraging people to use and develop

---

[19]See: McCrone et al. 2012; REN21 2013.

[20]See: Joshi et al. 2011.

clean energy technology. As these technologies are developed their costs will fall, in turn encouraging their use. Contrary to the objection that some make, developing cheaper clean energy partly depends on adopting policies to reduce emissions, whereas continuing to emit simply reinforces the dominance of fossil fuels in the global energy market. Whilst people should take efforts to develop clean energy technology, people should also be sceptical about the prospect of meeting the 2 °C target without taking emissions reductions soon.

In summary, there is a great deal of evidence suggesting that people should take immediate action to address climate change. These arguments relate to different aspects of climate change, yet they all point in the same direction: people cannot afford to delay taking action to mitigate our emissions and develop adaptation policies. Ordinarily, getting bogged down in procedural issues would mean losing out on the benefits of immediate collective action until some future point in time. With climate change things are very different. Prolonged inaction not only has severe consequences for the near future, it also makes it unlikely that catastrophic outcomes can be avoided in the long-term. This means that urgency should be given great importance here. The implications of this (which become clear in the next section) are that if reasonable disagreement prolongs action in the UNFCCC then this is a serious problem for meeting its overall goal and continued deliberation over intractable issues is not a viable option.

3.3.3 Consensual Agreement

So far, I've argued that the ultimate goal of the UNFCCC is to avoid dangerous climate change and that achieving this goal requires action that is both urgent and stringent. Now, I turn to an important condition under which the UNFCCC must achieve this goal, namely: consensual agreement. Consensual agreement means that the UNFCCC must command the voluntary consent and endorsements of all its participants.

The reasoning for this is that, under current international law, states cannot be forced to enter international agreements.[21] There are no sufficiently credible enforcement mechanisms for ensuring participation and compliance with multilateral treaties at this point in time.[22] This is because there is no global sovereign power that can force a state to participate in international treaties. There are also international norms that prevent states from forcing each other to participate in treaties. Even if this were not the case, then it is generally thought that there are insufficient mechanisms for encouraging participation and compliance at the

[21] Hurd 1999. For discussions relating this point to the UNFCCC, see: Wiener 1999, p. 769; Höhne et al. 2002, p. 10.

[22] There is a strong line of literature supporting this premise: Barrett and Stavins 2003; Vogler 2005; Chasek et al. 2006, p. 208.

international level. Whilst it might be the case that climate change eventually becomes so severe that it is imperative to force states to comply with the obligations of a multilateral treaty on climate change, credible sanctions and other means of enforcing compliance are not available at this point in time. This is because most enforcement mechanisms harm the enforcing state as much as they harm the offending state.[23] States are also reluctant to punish others, fearing that they may face reciprocal action.[24] Threats to impose economic sanctions or trade restrictions on another state therefore suffer from a credibility problem.[25] Whilst there are some limited enforcement mechanisms for ensuring compliance once a state has entered a treaty, these are relatively weak. Despite several instances of noncompliance, those who have refused to cooperate with the commitments of the UNFCCC have not yet faced any sort of punishment. Canada failed to meet its commitments under the Kyoto Protocol and walked away from the agreement unpunished whilst Japan, Russia and the US refused to sign up to a second commitment period without facing any penalties at all (Metz 2013).

The implication of this is that a necessary condition for cooperation on climate change is that actors comply with international institutions on a voluntary basis (Risse, M.A. 2004; Halsnæs et al. 2007, p. 127). Because the UNFCCC is a comprehensive agreement with universal membership of states on a global scale, the UNFCCC requires the cooperation of these actors at the global level. A further important point is that actors generally do not commit themselves to voluntary agreements unless they consider these agreements fair. This is often true for individuals, as well as at the international level.[26] That is, states are often unwilling to join an agreement if it thinks that the terms of the agreement are unfair, either to itself, or to other actors.[27] This means that a further condition of the UNFCCC is that all of its participants perceive it as fair. As I suggested in Chap. 2, this applies as much to the fairness of its procedures as it does to the fairness of its outcomes.

3.3.4 Necessary Criteria for the UNFCCC

To summarise the argument thus far, the ultimate goal of the UNFCCC is to limit atmospheric concentrations of greenhouse gases at a level that avoids dangerous climate change. This requires limiting global temperature increases to no more than

[23]For more on this point: Hoel 1992; Barrett 1994, 1998.

[24]I take this point from: Birnie 1988, p. 113.

[25]These measures are proposed in Aldy et al. (2001) and Nordhaus (1998).

[26]Thomas Franck argues that states obey international laws even when it is not in their interest to do so at least partly on account of the fairness of these laws (Franck 1995, p. 26). Peter Lawrence suggests that states may comply with an agreement if they feel that it is procedurally fair (Lawrence 2014, p. 16). For more on this point, see Barrett 2003.

[27]Barrett and Stavins 2003, p. 360.

2 °C above pre-industrial levels, which in turn requires urgent action. The fact that this requires urgent action carries additional importance that will become clear in the next section. Further to these points, the UNFCCC and its related institutions must gain the voluntary support of its participants. Since states are largely unwilling to support unfair institutions, it's also necessary that the international community perceives the UNFCCC as a fair institution.

This is where the relevance of reasonable disagreement becomes clear. As I argued in Chap. 2, there is reasonable disagreement about some of the substantive ends that a climate change institution should pursue. This means that agreement over the substantive ends of the institution is going to be difficult, if not wholly impossible. In fact, given the deadlock that has persisted in the negotiations of the UNFCCC for two decades, it seems fair to presume that this is unlikely to arise anytime soon, if ever at all. If avoiding climate change is something that requires great urgency, then this is obviously a problem.

Given these facts, I now argue that the parties of the UNFCCC might be able to reach mutually agreeable terms of cooperation through a fair procedure. In doing so, I show that procedural values are fundamentally important to the design of the UNFCCC.

3.4 Resolving Reasonable Disagreement

Given that there is reasonable disagreement in climate change, as well as a pressing need to implement action quickly, what can be done? This section considers several answers to this question, before advocating fair procedures as an element of one approach to achieving urgent action despite reasonable disagreement.

When faced with reasonable disagreement, one option is to continue to deliberate. After all, reasonable actors who disagree might ultimately come to change their minds. Deliberative democrats traditionally argue that conditions of cooperative deliberation in good faith can be expected to eliminate mistakes in reasoning to bring about consensus on an issue. For example, Gutmann and Thompson argue that deliberation can clarify a moral conflict by helping actors to see that it might be based on misunderstanding (Gutmann and Thompson 1996). The discussion of disagreement in Chap. 2 did not suggest that agreement was impossible, but rather that it was not forthcoming. By continuing to strive for agreement on a cooperative outcome, parties may reach consensus on an issue in the end.

But the downside of this approach is that it might take a very long time and may not yield an answer at all. The problem is that reasonable disagreement is likely to persist despite continued deliberation in good faith. Deliberative democracy works to bring about consensus by eliminating *unreasonable* claims. Here the concern is *reasonable* disagreement. Christopher McMahon goes as far as to argue that reasonable disagreement will persist no matter how long shared deliberation continues (McMahon 2009, p. 2). Furthermore, the evidence from climate negotiations so far gives no reason to think that states will come to agree on this issue anytime soon. Even where agreement initially seemed possible, two decades of negotiations have led to entrenched positions that delegations appear completely unwilling to back down from. The fair distribution of emission rights is just one highly contested issue that states are unwilling to compromise on. Whilst I don't suggest that agreement will never arise through continued discussion, I do suggest that reasonable disagreement is likely to lead to protracted and prolonged decision-making at the very least. It is therefore necessary to find a way of resolving disputes when deliberation doesn't bring about an outcome, or if time is a pressing matter.

If it isn't possible to achieve agreement through sustained deliberation, then a second option is to give up on the cooperative arrangement. One could forsake the cooperative arrangement altogether, either taking action alone or with others in a smaller group. This is something that's gaining increased prominence in the debates on climate change, as many have come to criticise the glacial progress of the UNFCCC. Increasingly, individual states are pursuing unilateral action on climate change at a domestic level.[28] Smaller groups of states are also entering bilateral, or 'minilateral' agreements with those who share common interests, achieving some of the benefits of cooperation whilst avoiding the pitfalls persistent disagreement.[29] These proposals represent strong candidates for alternative ways of achieving action on climate change that bypass the ineffective UNFCCC, and there is an on-going debate about whether international efforts to address climate change should abandon the UNFCCC in favour of these more productive agreements.[30] This is an extremely important debate over the future of the international climate change regime and one of the ultimate aims of this book is to contribute to this debate by advocating continued efforts within the UNFCCC. This is something that I fully discuss in Chap. 8. For the moment, I assume that, in accordance with its mandate, the UNFCCC should achieve universal membership of states on a global scale and that it cannot simply give up on this comprehensive approach in light of disagreement. This isn't to say that minilateral approaches aren't important features of the global effort on climate change and I later relax this assumption in Chap. 8.

A different approach is to force actors into accepting a proposal even if they disagree with it. One could impose threats, bribes or sanctions that enforce compliance with a certain agreement. After all, an actor may disagree with a particular proposal, but nevertheless consent to an agreement on the basis that not doing so would be prohibitively damaging. Coercive measures can provide a way of getting everyone to accept the terms of an agreement quickly, even in the face of reasonable disagreement. But, as I argued earlier, this seems unlikely in the current context of global politics. There is no global sovereign power that can impose obligations on states and current international law prohibits states from forcing

---

[28]Bottom-up processes involve voluntary mitigation pledges that are defined unilaterally (Bodansky 2012, p. 1).

[29]See: Christoff 2006; King et al. 2011, p. 19.

[30]For proponents, see: Prins and Rayner 2007; Prins et al. 2010.

others to participate in international treaties. Forcing actors to accept a proposal therefore seems an inappropriate way of dealing with reasonable disagreement in the UNFCCC.

A further option is to come up with a procedure to resolve disagreement. This could involve some kind of pragmatic procedure that helps actors to choose between different positions, even if it is not in a fair way. Here, I have in mind something similar to the procedures described by Jeremy Waldron during his discussion of decision-making by majority rule (Waldron 1999). These procedures include, for example, picking an option at random or deciding on an outcome through third party arbitration.[31] Such procedures aren't necessarily fair (I discuss what fair procedures might look like in the following section), and they only need to be sufficiently appealing so that they are acceptable to each actor involved. Such procedures would provide one way of achieving cooperative action even when there is reasonable disagreement about substantive ends.

Whilst enforced compliance and pragmatic procedures seem like possible solutions to the problem of reasonable disagreement, the problem, as Waldron notes, is that they fail to adequately respect actors.[32] As such, these methods of achieving consensus aren't fair. This is problematic for three reasons, the first of which relates to coercive enforcement, whilst the latter two relate to pragmatic procedures. The first problem is that coercive agreements are not sufficiently legitimate to create long-term and sustained cooperation. Successful action on climate change requires long-term action, and given the dynamic nature of global politics, it is unlikely that an institution will be able to adequately enforce coerced consensus in the very long-term. Second, it is unlikely that actors will comprehensively accept a procedure that is not fair. As I've already said, there is empirical evidence that shows that actors are less likely to accept an agreement that is unfair, whether the agreement is unfair for them, or if it is unfair for other actors.[33] A third problem is that agents must have reasons for accepting the outcome of a procedure when they disagree with the outcome itself. As it is, pragmatic procedures do not provide reasons for suspending one's own judgement for the sake of the outcome of a procedure. If actors are expected to support the outcome of a decision-making procedure, even if they disagree with the outcome, then there must be an additional reason for supporting it. Following Waldron, I take fairness to be a possible solution to this problem, and I now outline the argument why this is the case.

---

[31]Waldron suggests that two pragmatic procedures for decision-making might be tossing a coin or nominating one person to act as a leader (Waldron 1999, p. 113).

[32]Waldron 1999, p. 113. Note that continued deliberation and minilateralism may also fall short on this point.

[33]Thibaut and Walker 1975; Walker et al. 1979.

3.4.1 Procedural Fairness

So far I've suggested that fair procedures may be a way of dealing with reasonable disagreement in the UNFCCC. There are many accounts of procedural fairness in political theory and it's worth considering these various forms before outlining how fair procedures may help actors reach agreement in the UNFCCC. These forms of procedural fairness are differentiated by whether there is an independent criterion for analysing the substantive end of a decision-making process, and by whether a decision-making process exists to reach this end.

Perfect procedural justice involves a situation in which there is some ultimate, independent outcome that can be specified as just, or fair independently of the procedure in question, and there is a process that achieves this independent outcome. A common example used to illustrate perfect procedural justice is the division of a cake between two actors. In this case, it's general knowledge what a fair distribution is: each actor receives an equal portion. There is also a procedure that will bring about this outcome; if one actor is asked to cut the cake and the other has first choice over the available slices then each actor should receive an equal slice (assuming that each actor wants to maximize their share of the cake and is capable of cutting the cake in half).[34] In this instance, both the procedure and the outcome are known and are fair.

In contrast, imperfect procedural justice relates to situations in which there is an independent criterion for a fair outcome, but there is no known procedure for achieving this outcome. John Rawls argues that a trial by jury is one way of conceptualising this form of procedural justice.[35] In a trial by jury there is an independent criterion that we know to be just. That is, a just outcome is achieved when an accused person is convicted of a crime if and only if they are guilty of that crime. However, there is no perfect procedure for achieving this outcome in each and every instance. Criminal trials are complicated, costly and inaccurate. There are many unsolvable problems inherent in such processes and it is simply not possible to ensure that a guilty actor is correctly convicted of committing a crime in each and every case. This is, in part, because there are many competing values at stake in the design of a trial, and is difficult to satisfy all of these at the same time. But situations involving imperfect procedural justice also arise because it is simply not possible to design a procedure that can achieve a just outcome all of the time.

A third form that procedural justice can take is pure procedural justice.[36] Pure procedural justice is necessary when there is a situation in which there is no independent criterion for what justice requires, but there are procedures such that, if followed, can validate the outcome as just. The outcome achieved through a

---

[34]For discussion of perfect procedural justice, see: Kelsen 1955.

[35]See, Rawls 1971.

[36]For discussion of pure procedural justice see: Kelsen 1955; Rawls 1971, 1999; Nelson 1980; Arneson 2004.

fair procedure is itself fair by virtue of the process that generates it.[37] A fair democratic process is an example of pure procedural justice. For instance, if there is disagreement among members of a democratic polity about what constitutes a just distribution of resources, one might argue that the outcome of a democratic process is fair if it is reached through a democratically fair decision-making process. This requires adhering to certain democratic rules and procedures. Of course, this leaves open the question of what constitutes a democratically fair decision-making process in this instance. But this is not the point of the example given here (indeed, this question is the purpose of the rest of this book). Rather, the point is to illustrate that if it is possible to determine a sufficiently fair process for reaching decisions then such a process can render the outcomes of these decisions as just.

Chapter 2 demonstrated that there is significant, but reasonable disagreement about the fair distribution of emission rights in climate change institutions. As such, there is no independent criterion for specifying what a substantive just distribution is in this context. This is not to say that no such criterion exists. It might be the case that sufficient discussion and debate about substantive justice may lead parties to reach agreement on these issues. Alternatively, the fact that there is irresolvable disagreement about substantive justice does not preclude the existence of an objective outcome.[38] However, at this stage, it seems unlikely that actors in climate change institutions will reach agreement on the fair distribution of emission rights at any point in the near future. Given the urgency of achieving action on climate change, it seems reasonable to assume that either no independent criterion for substantive justice exists or, if it does exist, that it is (for all purposes) unknown and indeterminate. Given the absence of an independent criterion that specifies substantive justice, the argument of this chapter is that a procedural approach to the distribution of emission rights should be pursued instead. However, this does not mean that the outcome of a fair decision-making process is necessarily just, as is the case with pure procedural justice. Rather, the claim here is that a fair decision-making process generates an outcome that all can consider as sufficiently fair to be acceptable. The outcome of a fair decision-making process gains support as an acceptable outcome due to the fact that each actor believes that they are given sufficient respect in the decision-making process. As such, the proposal of this chapter represents a fourth conception of procedural justice, which draws on, but is distinct from, those types of procedural justice discussed above.[39]

[37]In this sense, the outcomes generated by a fair procedure are themselves fair. A related view might hold that, whilst there is disagreement on substantive justice, there is agreement on procedural justice, and that the outcomes generated by a fair process should be accepted as second best alternatives to persistent disagreement. For more on this, see: Arneson 2004.

[38]Jeremy Waldron claims that the fact that there is disagreement about an issue does not necessarily imply that there is no objectively right or correct outcome (Waldron 1999).

[39]This is similar to Thomas Christiano's account of 'moderate proceduralism', which recognises that democratic processes have intrinsic value whilst also placing limits on the substantive outcomes that these processes can achieve (Christiano 2008, p. 295).

Whilst this chapter focuses on the distribution of emission rights, this specific context can be expanded to the entire decision-making framework of the UNFCCC. The evident disagreement about the fair distribution of emission rights represents one example of the endemic disagreement that persists in negotiations to develop an institutional architecture for climate change more generally. This disagreement largely relates to the substantive ends that an agreement achieves. Here, the purpose of the remaining chapters of this book is to develop a set of principles that constitute a fair process for obtaining these outcomes. The argument is that, if these principles are implemented, the substantive outcomes reached through a decision-making process are acceptable to each party involved. The aim of the decision-making process is not to arrive at a substantively correct answer, but rather to a give each actor sufficient respect so that each can agree to the terms of cooperation of the institution.[40]

3.4.2 Resolving Disagreement Through Fair Procedures

In light of reasonable disagreement over urgent matters there is a strong case for a procedural approach to determining the outcomes that an institution should pursue. But it is not merely sufficient that we take a procedural approach; the procedure must also be perceived as fair. As I argued in Sect. 3.2, there are intrinsic reasons for doing this. But, as I also argued above, there are also instrumental reasons for fair procedures. This is because fair procedures take into account the moral worth of those who disagree. In doing so, fair procedures provide a way of responding to reasonable disagreement that recognises and respects difference rather than simply finding a pragmatic response to the problem of conflict. When arguing the case for majority rule, Waldron suggests that there are two ways in which this is the case (Waldron 1999, p. 111–4).

First, majority rule respects individuals because it does not rule out any single actor's point of view. No one's position is played down or ignored, and everyone's view is taken into account. Second, majority rule respects individuals by giving equal weight to each person's view in the decision-making process. The fact that majority rule respects individuals in these two ways gives each actor a reason for accepting the outcome of a procedure even if they disagree with the outcome itself. Whilst Waldron is arguing the case for majority rule, this argument can be expanded to other forms of decision-making more generally.[41] Here, I supplement Waldron's

---

[40]David Miller makes a similar argument about the role of deliberative democracy (Miller 2007). Thomas Christiano argues that democracy can provide a way of making decisions that treats each citizen equally when there are disagreements about justice and the common good in society (Christiano 2008, p. 75).

[41]Waldron notes that majority rule is not the only way of awarding sufficient respect in a decision-making process (Waldron 1999, p. 111).

argument with the claim that respecting actors in this way is also likely to bring about the necessary legitimacy and support for long-term and sustained cooperation. The upshot of this is that it isn't necessary to achieve consensus on what a just outcome on climate change is, provided that it's possible to identify a process that is sufficiently fair to bring about mutual acceptability.

Fair procedures, as I've set out here, capture some of the advantages of the various different approaches to resolving disagreement in the UNFCCC whilst avoiding some of their pitfalls. By bringing about agreement on a comprehensive scale, fair procedures achieve the level of participation that the UNFCCC sets out to achieve. They also provide an alternative to sustained deliberation over intractable issues, meaning that lengthy and deadlocked negotiations can be sidestepped. By appealing to fairness rather than pragmatism or force, this approach gains the long-term support necessary for sustained effective action on climate change.

This does not mean that procedural values should be prioritised over substantive ones, nor that there shouldn't be minimal substantive constraints on the outcomes that the UNFCCC brings about. Whilst this Chap. 2 showed that there is reasonable disagreement over some substantive values, it does not mean that there is reasonable disagreement about all substantive values in the design of the UNFCCC. Actors might agree on the minimal substantive values that should constrain the outcomes of procedures, whilst disagreeing about other important matters. But this chapter has emphasised the importance of considering procedural values in this context.

By providing an instrumental justification for pursuing procedural values, this account achieves the aims that were set out in Sect. 3.1. Achieving procedural fairness means operating in accordance with agreed rules of procedure. Determining exactly what the necessary requirements of procedural justice are for the UNFCCC is the subject matter of the remainder of this book. Before addressing this issue, it is worth addressing three objections to the arguments presented so far.

3.5 Procedural Values Appeal to Substantive Ends

One might claim that it is not possible to adopt a purely procedural approach to an issue such as the fair distribution of emission rights because, although actors recognise the importance of procedures, ultimately they are always concerned with some substantive end. This objection rests on two separate claims. First, that procedural fairness is not valuable in itself and fair procedures are only important insofar as they bring about some desired end.[42] This claim is supported by the fact that people often think that it is necessary to put limits on the types of outcomes that can arise from democratic procedures.[43] The second claim is that reasonableness

[42]Richard Arneson takes this view (Arneson 2004).

[43]Several authors suggest that there are certain constraints on the outcomes that multilateral institutions should bring about, for example: Buchanan and Keohane 2006.

involves a substantive judgement about the normative merit of various claims. If people are making substantive judgements about what claims are permissible, so the argument goes, then what they really care about is substantive ends.

To respond to the first claim, it is not the case that procedural values always collapse into substantive ones. As I argued in Sect. 3.2, procedural values are intrinsically important regardless of the outcomes they achieve. Further, just because people place some necessary substantive constraints on the outcomes that arise from decision-making procedures, it does not mean that they reject the importance of procedural values. Fair decision-making processes are important partly because they bring about mutual agreement. But this doesn't mean that people should agree to accept any outcome that arises from a procedure, nor does it mean that substantive values should be ignored altogether. The fact that there is reasonable disagreement over some substantive values does not mean that there is reasonable disagreement over all substantive values. Actors may reach consensus on certain minimal constraints that a procedure should achieve. Whilst disagreement over substantive ends necessitates fair decision-making processes, it does not negate the existence of substantive constraints on the outcomes of these processes.

Turning to the second claim, it is not the case that judging the reasonableness of a claim necessarily involves a judgement about its substantive merit. I've suggested that reasonable actors should respect the reasonable claims of others. But I've defined reasonable actors as those who come together to cooperate fairly and in good faith. Procedural fairness requires that actors engage with one another under conditions of tolerance and mutual accommodation.[44] This suggests that people should respect the claims of others, on the basis of their reasonable behaviour, rather than on the substantive merit of their claims.

3.5.1 Disagreement About Procedural Values

A different objection is that there may be significant reasonable disagreement about what a fair decision-making process is.[45] Given the extent of disagreement over the substantive ends of climate change it seems reasonable to propose that there is going to be just as much disagreement over procedural values. There are many areas in which disagreement can arise in the design of procedures and these issues might be highly contested. In fact, the UNFCCC is yet to adopt formal voting rules for making decisions for the very reason that parties have been unable to agree on this issue. Therefore, some might claim that the procedural method defined here is unsuitable

[44]Here I have in mind something along the lines of Gutmann and Thompsons account of reciprocity (Gutmann and Thompson 1996).

[45]For instance, Ronald Dworkin argues that reasonable citizens may disagree about what democracy requires (Dworkin 1996, p. 34).

for achieving a fair agreement because it is subject to the same disagreement that gave it merit in the first place.[46]

There are at least three responses to this objection. One might hold that there is, in fact, a greater degree of agreement on procedural issues than there is about substantive ones. For example, Gutmann and Thompson argue that procedural values are often invoked as higher order principles that transcend disagreement on substantive issues (Gutmann and Thompson 1990, p. 64). In relation to climate change, substantive issues often involve distributing costs and benefits, which is a matter that is highly politicised with intractable positions. Procedural values, on the other hand, often invoke more abstract concepts that do not relate to such contentious and politically charged issues. Further, climate change is characterised by extreme uncertainty, which makes it difficult for actors to reach agreement on the substantive outcomes of climate change agreements. Some authors argue that, when outcomes are uncertain, actors put an emphasis on the quality of the procedure, over substantive outcomes (Toth 1999, p. 2; Foti et al. 2008). For this reason, procedural fairness can be an important precondition for reaching agreement in the UNFCCC. The hope is that, whilst there may be some disagreement on procedural values in climate change, this is not as intractable as that concerning substantive issues.

Further, parties may be more amenable to agreement on a procedural issues precisely because there is so much disagreement over the substantive ends in the UNFCCC. Climate change negotiations take place among state delegates who are accountable to the domestic governments that they represent. Many positions in climate change negotiations have become non-negotiable, or intractable, because any form of concession or compromise is seen as a sign of failure or defeat. These positions have become so entrenched that it is simply not possible to reach agreement in practice, even if it is possible to specify a fair agreement in theory. In this case, a procedural approach might provide an alternative way around entrenched views that negotiators cannot back down from. By respecting difference, rather than proposing a substantive answer, a procedural approach can resolve the problem of entrenched disagreement, rather than creating further disagreement about fair process. This is an empirical matter about the acceptability of different types of claims. However, given that I'm concerned with the way that actors negotiate in climate change, this remains an important point.

But regardless of what one might think about the plausibility of reaching agreement on procedural issues, it remains the case that this area is so far unaddressed in the literature on climate change. There may be as much inherent disagreement about procedural values as there is about substantive values in climate change, but there is (to date) no analysis of procedural values in this context. The purpose of the remainder of this book is to explore these issues further and to consider their feasibility and acceptability as principles of justice. If, on further investigation, it is evident that there is fundamental and prohibitive disagreement in issues of process,

[46]For discussions of disagreement over procedural values see: Waldron 1999; Karlsson 2008; Mansbridge et al. 2010.

then it may be necessary to undertake an alternative approach to achieving justice in climate change. But, it would be presumptuous to rule out a procedural approach without any evidence of actual disagreement.

3.5.2 Why Should People Accept the Outcome of a Procedure?

A further objection to a procedural approach questions why we should expect actors to accept the result of a procedure if they disagree with the outcome. As Jeremy Waldron points out, some people might object to a procedural approach on the basis that it asks them to prioritise procedure above substance (Waldron 1999, p. 160–1). Given the significance of forsaking one's own view for the outcome of a procedure, so the argument goes, it's necessary to have a justification for prioritising procedure above substance. If justice is the most important virtue of institutions, then surely procedures shouldn't take precedence.

Waldron argues that this objection rests on a misunderstanding of what he calls 'the dimensions of political importance'. Just because people accept the outcome of a procedure above their own view on an issue does not mean that they place procedural considerations above substantive ideas of justice. Rather, by adopting the outcome of a procedure they accept that there is a need to reach common agreement in the face of reasonable disagreement. By taking a procedural approach people do not prioritise this above justice, but accept the need to find agreement given their disagreement. Given the urgency and importance of mitigating climate change, there might be hope that actors can come to see the validity of others' reasonable positions in the UNFCCC, provided that the final outcome is determined in a fair way.

It's also worth recognising that there are limits to what procedures can achieve. Fair procedures produce a result that everyone might be prepared to accept, even if they do not agree with the outcome wholeheartedly. I'm proposing that if there is initial disagreement about something then a fair process might lead us to a result that all can accept. But this might be impossible if the initial disagreement is severe and deep. People may be unwilling to accept the outcome of a procedure, even if it comes about in a fair way. Whilst this is something that should be kept in mind, it doesn't preclude taking a procedural approach. It is important to consider the design of fair procedures whilst accepting that this might not always lead to a result that every actor is prepared to accept.

3.6 Conclusion

This chapter has made the central claims that procedural fairness is both intrinsically, and instrumentally valuable in the UNFCCC. It is intrinsically valuable because actors value the fact that processes are, in fact, fair. It is instrumentally valuable because it allows parties to reach agreement on issues on which there is reasonable disagreement, where reasonable disagreement implies that there are different reasonable positions on an issue and no single option dominates any other. Given that the UNFCCC requires action which is both urgent and which commands the consensual cooperation of all its parties, procedural values gain addition importance in the design of the UNFCCC. The remainder of this book considers what a fair procedure is, and what we can expect from reasonable actors.

References

Agarwal, A., and S. Narain. 1991. *Global warming in an unequal world, a case of environmental colonialism*. New Delhi: Centre for Science and Environment.

Aldy, J.E., P.R. Orszag, et al. 2001. *Climate change: An agenda for global collective action*. Pew: Center on Global Climate Change.

Allen, M.R., D.J. Frame, et al. 2009a. Commentary: The exit strategy. *Nature Reports Climate Change* 3: 56–58.

Allen, M.R., D.J. Frame, et al. 2009b. Warming caused by cumulative carbon emissions towards the trillionth tonne. *Nature* 458: 1163–1166.

Alley, R.B., J. Marotzke, et al. 2005. Abrupt climate change. *Science* 299: 2005–2010.

Arneson, R.J. 2004. Democracy is not intrinsically just. In *Justice and democracy*, ed. K. Dowding, R.E. Goodin and C. Pateman. Cambridge: Cambridge University Press.

Bäckstrand, K. 2010. Democratizing global governance of climate change after Copenhagen. In *Oxford handbook on climate change and society*, ed. J. Dryzek, R.B. Norgaard, and D. Schlosberg. Oxford: Oxford University Press.

Barrett, S. 1994. Self-enforcing international environmental agreements. *Oxford Economic Papers* 46: 878–894.

Barrett, S. 1998. The political economy of the Kyoto protocol. *Oxford Review of Economic Policy* 14: 20–39.

Barrett, S. 2003. *Environment and statecraft*. Oxford: Oxford University Press.

Barrett, S., and R.N. Stavins. 2003. Increasing participation and compliance in international climate change agreement. *International Environmental Agreements: Politics, Law and Economics* 3: 349–376.

Beitz, C.R. 1989. *Political equality: An essay in democratic theory*. Princeton: Princeton University Press.

Birnie, P. 1988. International law and solving conflicts. In *International environmental diplomacy: The management and resolution of transfrontier environmental problems*, ed. J.E. Carroll. Cambridge: Cambridge University Press.

Bodansky, D. 2012. *The Durban platform: Issues and options for a 2015 agreement*. Centre for Climate and Energy Solutions

Bodansky, D., and L. Rajamani. 2013. Evolution and governance architecture. In *International relations and global climate change*, ed. D. Sprinz and U. Luterbacher. Cambridge, MA/London: MIT Press.

Bruce, J.P., L. Hoesung, et al. 1995. Climate change 1995: Economic and social dimension of climate change. In *Contribution of Working Group III to the Second Assessment Report of the Intergovernmental Panel on Climate Change*. Cambridge: Cambridge University Press.

Buchanan, A.E., and R.O. Keohane. 2006. The legitimacy of global governance institutions. *Ethics and International Affairs* 20(4): 412.

Chasek, P., and L. Rajamani. 2003. Steps toward enhanced parity: Negotiating capacities and strategies of developing countries. In *Providing public goods: Managing globalization*, ed. I. Kaul, P. Conceição, K. Goulven, and F. Mendoza. Oxford: Oxford University Press.

Chasek, P.S., D.L. Downie, et al. 2006. *Global environmental politics*. Boulder: Westview Press.
Christiano, T. 1996. *The rule of the many: Fundamental issues in democratic theory*. Boulder/Oxford: Westview Press.
Christiano, T. 2008. *The constitution of equality: Democratic authority and its limits*. Oxford: Oxford University Press.
Christoff, P. 2006. Post-Kyoto? Post-bush? Towards an effective climate coalition of the willing. *International Affairs* 82(5): 831–860.
den Elzen, M. 2010. The emissions gap report. UNEP. http://www.unep.org/publications/ebooks/emissionsgapreport/.
den Elzen, M., M. Meinshausen, et al. 2006. Multi-gas emission envelopes to meet greenhouse gas concentration targets: Costs versus certainty of limiting temperature increase. *Global Environmental Change* 17: 260.
Dirix, J., W. Peeters, et al. 2013. Strengthening bottom-up and top-down climate governance. *Climate Policy* 13(3): 363–383.
Dubash, N.K. 2009. Copenhagen: Climate of mistrust. *Economic & Political Weekly* XLIV(52): 8–11.
Dworkin, R. 1996. *Freedom's law: The moral reading of the American constitution*. Oxford: Oxford University Press.
Eckersley, R. 2012. Moving forward in the climate negotiations: Multilateralism or minilateralism? *Global Environmental Politics* 12(2): 24–42.
Foti, E., L. de Silva, et al. 2008. *Voice and choice: Opening the door to environmental democracy*. Washington, DC: Word Resources Institute.
Frolicher, T.L., and F. Joos. 2010. Reversible and irreversible impacts of greenhouse gas emissions in multi-century projections with the NCAR global coupled carbon cycle-climate model. *Climate Dynamics* 39: 1439–1459.
Gardiner, S. 2011. *A perfect moral storm: The ethical tragedy of climate change*. New York: Oxford University Press.
Global Commission on the Economy and Climate. 2014. The new climate economy. Available at: http://newclimateeconomy.net/content/global-commission
Gupta, J. 2000. *"On Behalf of My Delegation . . . " A survival guide for developing country climate negotiators*. Published jointly by the Center for Sustainable Development of the Americas and the International Institute for Sustainable Development.
Gupta, S., D.A. Tirpak, et al. 2007. Policies, instruments and co-operative arrangements. In *Climate change 2007: Mitigation. Contribution of working group III to the fourth assessment report of the Intergovernmental Panel on Climate Change*, ed. B. Metz, O.R. Davidson, P.R. Bosch, R. Dave and L.A. Meyer. Cambridge: Cambridge University Press.
Gutmann, A., and D. Thompson. 1990. Moral conflict and political consensus. *Ethics* 101(1): 64–88.
Gutmann, A., and D. Thompson. 1996. *Democracy and disagreement*. Cambridge, MA: Harvard University Press.
Halsnæs, K., P.R. Shukla, et al. 2007. Framing issues. In *Climate change 2007: Mitigation. Contribution of working group III to the fourth assessment report of the Inter-governmental Panel on Climate Change*, ed. B. Metz, O.R. Davidson, P.R. Bosch, R. Dave and L.A. Meyer (eds.). Cambridge: Cambridge University Press.
Hare, B., and M. Meinshausen. 2006. How much warming are we committed to and how much can be avoided? *Climate Change* 75: 1–2.
Heyward, M. 2007. Equity and international climate change negotiations: A matter of perspective. *Climate Policy* 7(6): 518–534.
Heyward, C. 2010. Environment and cultural identity: Towards a new dimension of climate justice. Doctoral thesis; the Department of Politics and International Relations, at the University of Oxford.
Hoel, M. 1992. International environmental conventions: The case of uniform reductions of emissions. *Environmental and Resource Economics* 2: 141–159.

Höhne, N., C. Galleguillos, et al. 2002. *Evolution of commitments under the UNFCCC: Involving newly industrialized economies and developing countries*. The German Federal Environmental Agency (Umweltbundesamt).
The Human Rights Council. 2008. *Human rights and climate change*. http://www.ohchr.org/EN/Issues/HRAndClimateChange/Pages/HRClimateChangeIndex.aspx
Hurd, I. 1999. Legitimacy and authority in international politics. *International Organization* 53(2): 379–408.
IEA. 2011. *World energy outlook 2011: Executive summary*. Paris: International Energy Agency. http://www.worldenergyoutlook.org/media/weowebsite/2011/executive_summary.pdf.
IISD. 2010. Earth negotiations bulletin: Summary of the Cancun climate change conference. *Earth Negotiations Bulletin* 12(498): 1–30.
IPCC. 2007. Fourth assessment report: Climate change 2007. In *Contribution of working groups I, II and III to the fourth assessment report of the Intergovernmental Panel on Climate Change*, ed. Core Writing Team, Pachauri, R.K. and Reisinger, A. Geneva: IPCC.
IPCC. 2013. *Climate change 2013: The physical science basis: Summary for policymakers*. Working Group I Contribution to the IPCC Fifth Assessment Report. Intergovernmental Panel on Climate Change.
Joshi, M., E. Hawkins, et al. 2011. Projections of when temperature change will exceed 2 °C above pre-industrial levels. *Nature Climate Change* 1: 407–412.
Karlsson, J. 2008. Democrats without borders: A critique of transnational democracy. In *Gothenburg studies in politics*, ed. Bo Rothstein. Gothenburg: Department of Political Science, University of Gothenburg.
Karlsson-Vinkhuyzen, S.I., and J. McGee. 2013. Legitimacy in an era of fragmentation: The case of global climate governance. *Global Environmental Politics* 13(3): 56–78.
Kelsen, H. 1955. Foundations of democracy. *Ethics* 66(1): 1–101.
King, D., K. Richards, et al. 2011. *International climate change negotiations: Key lessons and next steps*. Oxford: Smith School of Enterprise and the Environment, The University of Oxford.
Lange et al. 2007. On the importance of equity in international climate policy: An empirical analysis. *Energy Economics* 29(3): 545–562
Lawrence, P. 2014. *Justice for future generations*. Cheltenham: Edward Elgar.
Lenton, T.M. 2011. Early warning of climate tipping points. *Nature Climate Change* 1: 201–209.
Lomborg, B. 2001. *The sceptical environmentalist*. Cambridge: Cambridge University Press.
Mansbridge, J., et al. 2010. The place of self-interest and the role of power in deliberative democracy. *Journal of Political Philosophy* 18(1): 64–100.
McCrone, A., E. Usher, V. Sonntag-O'Brien, U. Moslener, and C. Grüning. 2012. *Global Trends in Renewable Energy Investment 2012*. Frankfurt School UNEP Collaborating Centre for Climate & Sustainable Energy Finance.
McMahon, C. 2009. *Resonable disagreement*. Cambridge: Cambridge University Press.
MEF. 2009. *Declaration of the leaders*. The Major Economies Forum on Energy and Climate. http://www.whitehouse.gov/the_press_office/Declaration-of-the-Leaders-the-Major-Economies-Forum-on-Energy-and-Climate
Meinshausen, M., N. Meinshausen, et al. 2009. Greenhouse-gas emission targets for limiting global warming to 2 °C. *Nature* 458(7242): 1158.
Metz, B. 2013. The legacy of the Kyoto protocol: A view from the policy world. *WIREs Climate Change* 4: 51–158.
Miller, D. 2007. Deliberative democracy and social choice. *Political Studies* 40: 54–67.
Nelson, W. 1980. The very idea of pure procedural justice. *Ethics* 90(4): 502–511.
Nordhaus, W.D. 1998. *Economics and policy issues in climate change*. Washington, DC: Resources for the Future.
Oliver, J., G. Janssens-Maenhout, and J.A.H.W. Peters. 2012. *Trends in global CO_2emissions; 2012 Report*. The Hague/Ispra: PBL Netherlands Environmental Assessment Agency/Joint Research Centre.
Patz, J.A., et al. 2005. Impact of regional climate change on human health. *Nature* 438(17): 310–317.

Peters, G.P., R.M. Andrew, et al. 2013. The challenge to keep global warming below 2 °C. *Nature Climate Change* 3(1): 4–6.
Prins, G., and S. Rayner. 2007. *The wrong trousers: Radically rethinking climate policy*. Joint Discussion Paper of the James Martin Institute for Science and Civilization, University of Oxford and the MacKinder Centre for the Study of Long-Wave Events, London School of Economics.
Prins, G., I. Galiana, et al. 2010. *The Hartwell paper: A new direction for climate policy after the crash of 2009*. Oxford: Institute for science, Innovation and Society, University of Oxford.
Rawls, J. 1971. *A theory of justice*. Cambridge, MA: Harvard University Press.
Rawls, J. 1999. *The law of peoples*. Cambridge, MA/London: Harvard University Press.
REN21. 2013. *Renewables 2013*. Global Status Report. Paris: REN21 Secretariat.
Risse, M. 2004. What we wwe to the global poor. *Journal of Ethics* 9: 81–117.
Rogelj, J., et al. 2011. Emission pathways consistent with a 2 °C global temperature limit. *Nature Climate Change* 1: 413–418.
Schneider, S.H., and J. Lane. 2006. Dangers and thresholds in climate change and the implications for justice. In *Fairness in adaptation to climate change*, ed. N.W. Adger, J. Paavola, S. Huq and M.J. Mace. Cambridge, MA: MIT Press.
Solomon, S., G.-K. Plattner, et al. 2009. Irreversible climate change due to carbon dioxide emissions. *Proceedings of the National Academy of Science of the United States of America* 106(6): 1704–1709.
Steinacher, M., F. Joos, et al. 2013. Allowable carbon emissions lowered by multiple climate targets. *Nature* 499: 197–201.
Stern, N. 2014. The costs of delaying action to stem climate change. Available at: http://www.whitehouse.gov/sites/default/files/docs/the_cost_of_delaying_action_to_stem_climate_change.pdf
Stocker, T.F. 2013. The closing door of climate targets. *Science* 339: 280–282.
Thibaut, J., and L. Walker. 1975. *Procedural justice: A psychological analysis*. New York: Wiley.
Toth, F.L. 1999. *Fair weather? equity concerns in climate change*. London: Earthscan.
UNFCCC. 1992. *United Nations framework convention on climate change*. Convention Text.
UNFCCC. 2009. *The Copenhagen accord*. Report of the conference of the parties on its fifteenth session, held in Copenhagen from 7 to 19 December 2009.
UNFCCC. 2010. *The Cancun agreements*. Report of the conference of the parties on its sixteenth session, held in Cancun from 29 November to 10 December 2010.
UNFCCC. 2011. *The Durban platform*. Draft decision -/CP.17 Establishment of an Ad Hoc Working Group on the Durban Platform for Enhanced Action.
UNFCCC. 2014 *Lima call for climate action*. United Nations Framework Convention on Climate Change.
Unruh, G.C. 2000. Understanding Carbon Lock-in. *Energy Policy* 28: 817–830.
Vogler, J. 2005. In defence of international environmental cooperation. In *The state and the global ecological crisis*, ed. J. Barry and R. Eckersley. Cambridge, MA: MIT Press.
Voorhar, R., and L. Myllyvirta. 2012. *Point of no return: The massive climate threats we must avoid*. Amsterdam: International Greenpeace.
Waldron, J. 1999. *Law and disagreement*. Oxford/New York: Oxford University Press.
Walker, L., E.A. Lind, and J. Thibaut. 1979. The relation between procedural and distributive justice. *Virginia Law Review* 65(8): 1401–1420.
Weitzman, M.L. 2009. On modeling and interpreting the economics of catastrophic climate change. *Review of Economics and Statistics* 91(1): 1–19.
Wiener, J.B. 1999. On the political economy of global environmental regulation. *Georgetown Law Journal* 87: 749–794.

Chapter 4
Getting a Seat at the Table: Fair Participation in the UNFCCC

4.1 Introduction

Having made a case for fair procedures, Chap. 4 now turns to the question of what procedural fairness requires in the UNFCCC by considering who should participate in its decisions. Procedural justice is often understood as requiring that all those who are affected by the outcome of a decision should have some say in the decision making process (the All Affected Principle). Yet, there are many objections to this approach, there are also many other principles of fair participation to consider, and it is not immediately apparent that this principle should be applied in the UNFCCC. Furthermore, increasing the number of participants in a decision is often detrimental to the ability to reach agreement on an issue. In this chapter, I discuss the merit of the All Affected Principle and consider how fair participation can be achieved in the UNFCCC. I analyse several alternative principles for fair inclusion in the decisions of the UNFCCC and argue that fair processes are those provide representation to states on a global scale. I then consider what procedural rules are required in order to achieve this in the UNFCCC.

The decisions made in the UNFCCC affect people in a profound way on a worldwide scale. Climate change is a global phenomenon, which means that decisions concerning the mitigation of climate change have implications for people everywhere. Other decisions, such as those governing adaptation measures, or the distribution of climate finance, may not affect people globally, but still affect large numbers of people in a considerable way. It's often thought that people who are affected by a decision should have some say in the way that it is made. For this reason, the UNFCCC has traditionally sought a high level of participation in its

L. Tomlinson, *Procedural Justice in the United Nations Framework Convention on Climate Change*, DOI 10.1007/978-3-319-17184-5_4

decisions, both through the universal participation of its member states, as well as through the involvement of many civil society actors.[1]

The COP represents the main decision-making forum of the UNFCCC. Currently, decisions in the UNFCCC are formally adopted by its constituent member states when consensus is reached. This gives an equal say in the UNFCCC's decisions to each member state. Member state delegations are in privileged in the sense that they are the sole actors that have a right to oppose a decision, or to have a vote. Delegations are also privileged in the sense that they have other procedural rights, such as the right to make formal interventions in COP negotiations. Many other, 'non-state' actors (NSAs) also participate in these decisions, although they do not hold the same rights as states. These actors participate in these decisions in more informal ways, influencing decisions by providing information and assistance to state delegations, as well as making interventions on the request of the COP Chair.[2] The fact that these actors open a window into the COP from the wider world and provide a voice that otherwise goes unrepresented gives the UNFCCC added legitimacy. The value that these actors bring to discussions has led the UNFCCC to adopt an inclusive policy towards participation in COP negotiations.[3] The list of observer organisations, which have the right to access the COP negotiation forum and informally participate in its decisions, has increased exponentially since the early 1990s. At COP15 in Copenhagen there were 37,000 registered participants, including 10,500 government delegates, 13,500 observers from civil society, and 3,000 journalists (Dimitrov 2010).

But the procedural deadlock experienced in the UNFCCC since 1992 raises questions about the suitability of this approach. The number of actors that participate in multilateral decisions has important implications for how easy it is to make decisions and reach agreement. The more actors that there are in a decision-making process, the harder it is to aggregate interests and come to an agreement. More views are on the table, and more voices demand to be heard. This has led some to claim that open-ended institutional membership combined with decision-making by consensus has contributed to the stalemate within the UNFCCC. A number of commentators have suggested that the UN's principles of unanimity and inclusiveness contribute to deadlock in the negotiations of the UNFCCC.[4]

In response to this, some have sought to make decisions amongst smaller, more exclusive groups for decisions in the UNFCCC. Whilst this avoids the procedural problems associated with making decisions in large groups, it also comes under heavy criticism for its lack of procedural fairness. During the closing stages of

---

[1]See: UNFCCC 1992, Articles 4.1, 6 and 7. For discussion on the role of NSAs in COP negotiations, see: Gupta et al. 2007.

[2]For discussion of the various roles that NSAs play in climate institutions see: Jagers and Stripple 2003; Bulkeley and Newell 2010; Kravchenko 2010.

[3]UNFCCC 1992 Art. 4.1(i).

[4]For discussions of deadlock in climate change negotiations, see: Victor 2006; Haas 2008; Dombrowski 2010, p. 413.

COP15 in Copenhagen, delegates from four countries met in private to negotiate an agreement outside the main negotiation forum.[5] This group later declared that a global deal had been achieved, sparking outrage from the broader COP delegation who hadn't been informed of the agreement and were unaware of the discussions that had taken place (Rapp et al. 2010). Subsequently, Venezuela, Cuba, Nicaragua and Bolivia later renounced the agreement on the grounds that they had been excluded from important meetings (Bodansky and Rajamani 2013, p. 12). Similarly, the UNFCCC Climate Change Conference in Bangkok (2009) also drew heavy criticism after media representatives and other NSAs were excluded from participating in the negotiation sessions (Vidal 2009).

Whilst this raises questions about whether or not participation in COP negotiations should be limited on the grounds that it over-burdens decisions, there are also reasons for thinking that participation should be limited where actors are affected by the outcomes of these decisions to different degrees. At the 1997 COP negotiations in Kyoto, all of the UNFCCC constituent states participated in decisions concerning the negotiation outcome document, even though only a subset of these actors were actually subject to its provisions. This suggests that participation may important for those who aren't formally bound by a decision, but are the nonetheless significantly affected by it. These concerns extend beyond state delegations; they also relate to NSAs.

The decisions made within the UNFCCC clearly have considerable implications, not just for those who participate in the COP, but also for people globally. This raises questions about who should participate in the decisions of the UNFCCC as a matter of fairness, and how this should be balanced against the need to make decisions effectively. Much of the literature on democracy and multilateralism advocates greater participation of those who are externally affected by the outcomes of international institutions, through increased accountability, representation, or transparency in the processes that bring these outcomes about.[6] This is also something that's been embedded in the formal rules of other multilateral environmental agreements.[7] Those who advocate democratic inclusion in this way typically make two important assumptions. First, fairness demands that those who are affected by the outcomes of an institution are included in its decision-making processes (the so-called 'All Affected Principle'). Second, deciding who should participate is a matter of what's feasible in practice. Whilst some suggest that the demands of democratic inclusion require ideal responses to global governance such as electoral accountability on a global scale (Falk and Strauss 2000), others suggest that such approaches are unachievable in the current system of world politics, and that greater inclusion should be achieved through democratic processes that are neither accountable nor directly representative (Dryzek and Stevenson 2011, 2012a).

[5]For discussion of this, see: Dimitrov 2010; Rapp et al. 2010.

[6]For example: Archibugi et al. 1998; Palerm 1999; Bäckstrand 2006, 2010a, b.

[7]UNCED 1992, Principle 10; UNECE 1998, Article 6.

In this chapter, I question the former of these assumptions. I argue that those who are affected by the outcome of a decision do not, by virtue of the fact that they are affected, have a right to a say in the way that it is made. I claim instead that a more accurate understanding of procedural fairness is that those who are coerced by the outcomes of a decision, as well as those in whose name a decision is made, should be included in its decision-making processes. I go on to argue that those whose interests are affected by a decision should have a right to be heard in a decision, although this does not entail a right to participate in the sense of having a say in the way that it's made. With this in mind, the next section considers how we might start to think about fair participation in general contexts. In what follows, I consider some of the most frequently discussed principles for thinking about how to determine a democratic constituency.[8]

4.1.1 In One's Name

One common way of determining the demos is to say that participation is required if a decision is made *in an actor's name* (Beerbohm 2012). That is to say, if a decision is made on an actor's behalf, or if a decision is attributed to an actor, then that actor should have some say in the way that the decision is made. This carries strong intuitive appeal. People typically think that it only makes sense to attribute a decision to an actor if that person has had some say in the decision-making process. In a domestic context people have a right to participate in a decision-making process if a law is made in their name, regardless of whether they are actually affected by it in any real sense.

For sure, an agent may claim to act on behalf of an individual, in the sense that it takes the individuals interests into account and make the best decision for them, without actually giving that individual a say in the way that the decision is made. But it's wrong to say that it's fair to make a decision on someone's behalf whilst excluding that person from the decision-making process. The fact that a decision is made in someone's name means that, in some sense, that person should authorise and accept that decision. Carrying out a decision in someone's name without giving him or her a say in how it is made contradicts this.

It is difficult to think of cases where this doesn't seem to be an appropriate principle of fairness. If a decision is made in someone's name, then it often seems unfair if that actor is not included in the decision. It might be fair to make a decision on an actor's behalf if, for example, that actor has elected someone to make a decision for them. Alternatively, it might seem reasonable to make a decision in someone's name if he or she is incapable of making decisions that represent his or her best interests. But these are exceptional cases. In the first instance, the fact that an actor endorses a representative agent to make a decision is an act of participation

[8] In what follows, I draw from the discussions in Miller (2009) and Goodin (2007).

itself, and doesn't contradict our idea of what's fair. In the second case people can act on behalf of someone else if that person's particularly poor at making decisions. Provided that someone possesses some minimum capability to make a decision, there's no reason for excluding him or her from a decision. Assuming that the majority of the actors in a multilateral context are capable of making good decisions that represent their best interests, this second point isn't problematic for our idea of what's fair either. So the idea that people should be able to participate in decisions that are made in their name seems like plausible principle of justice for democratic inclusion and should govern at least some of our thoughts about who should participate in the decisions of the UNFCCC.

So who does the UNFCCC make decisions on behalf of? The UNFCCC aims to have universal membership amongst states globally. The COP of the UNFCCC makes decisions in the name of its member states. That is, the COP is a decision-making group that consists of the delegations of the UNFCCC signatory states. In this respect, we might think that fairness requires that these member states should have a say in the way that the UNFCCC makes its decisions. But this clearly isn't the whole story here. If fairness requires that those in whose name a decision is made should have a say in the UNFCCC, then surely this extends to those who make decisions within the UNFCCC as well. States act on behalf of many types of actors aside from just those in their constituencies. Many states aren't democratic accountable. In light of the many repressive and authoritarian regimes that exist in the world, it becomes hard to justify the claim that every state acts on behalf of its constituent citizens. So it's also important to think about how to give those in whose name the UNFCCC acts, but who are not included in its decisions. This is something that I take up in the final section of this chapter where I make several proposals for the reform of the UNFCCC. For the moment, its sufficient to recognise that this principle should play some part in our thinking of fair participation in the UNFCCC. Acknowledging this, I now turn to several other claims for fairness.

4.1.2 The All Affected Principle

The most prominent claim for fair participation is that all those who are affected by the outcome of a decision should have some say in how it is made.[9] This is known as the All Affected Principle (AAP),[10] and it has gained considerable traction in multilateral environmental politics, where it has become a defining feature of many international environmental agreements.[11]

[9]For discussion, see: Banuri et al. 1995, p. 86; Held 2004; Shelton 2007, p. 660.

[10]For support of the AAP, see: Dahl 1975; Lijphart 1984; Goodin 2007. For discussion of the AAP, see: Whelan 1983; Arrhenius 2005, p. 6; Agné 2006.

[11]Most notably the Aarhus Convention (UNECE 1998) the Rio Declaration on Environment and Development (UNCED 1992).

At first glance, the AAP appears an appealing candidate for determining who should participate in decisions as a matter of fairness. Assuming that interests are important, and that people are generally the best representatives of their own interests, then it seems reasonable that agents should have some say in the decisions that affect them.[12] The AAP also fits with many of our intuitions about who should participate in a decision. If a government is deciding whether or not to build infrastructure that results in the forced relocation of an indigenous group, then it's important that this group has a say in the decision-making process. Likewise, people feel that a local community should have some say in whether or not a large supermarket is built in a local town, or whether a polluting power plant is built nearby, or whether the Government licenses shale gas extraction.

But on further thought, the AAP doesn't seem so appropriate in each and every situation and there are many who argue against this principle.[13] Before considering why we might reject the AAP on theoretical grounds, it's worth pointing out some of the considerable problems that arise with implementing this principle in practice. Foremost, whilst the AAP implies that all those who are affected by a decision should have some say in the way that it's made, it says nothing about what it means to be affected, nor what sort of say this entails. There are many conflicting accounts of what it means to be affected in a way would warrants inclusion in a decision and what form this inclusion should take. I might claim that a decision to destroy an area of natural beauty in another part of the world affects me in some way, even though I'm not actually affected by the decision in any physical sense. I might argue that I have a strong claim to a say in the way that the decision is made, or I might only claim that my views about this matter should be respected in the decision. Whatever one thinks about the merit of the AAP in theory, a more refined account of what affectedness means is clearly needed before it can be adopted as a principle for policy reform.

Assuming that people can reach some understanding about what it means to be affected in a way that's relevant for participation, in certain situations a second problem arises. For some decisions, the decision-making process itself determines which actors are actually affected by its outcome.[14] If a decision is made about where to site a noisy wind turbine, then the outcome of the decision determines who is actually affected by the noise of this turbine. But the identification of those who are actually affected by this decision can't be determined until the decision is made. Adopting an actually affected interpretation of the AAP appears incoherent here, because the decision determines who is affected. In addition to this, the actually affected view takes the status quo as the baseline from which to judge whether an actor is affected by a decision. But, as Robert Goodin rightly argues, if a decision-

[12]Peter Lawrence advocates this principle for the UNFCCC on the grounds that it has the greatest chance of bringing about outcomes that are substantively just (Lawrence 2014, p. 188).

[13]For criticism of the AAP see: Karlsson 2006; Miller 2009; Schaffer 2012.

[14]Toth et al. 2001, p. 669; Goodin 2007, p. 53.

making body chooses one option from a set of many, then an actor can be affected in the sense that he or she would have been in a different situation, had that option not been chosen (Goodin 2007, p. 53). If an employer is deciding whom to employ from a group of several potential candidates, then each candidate is affected by her decision, even though the situation of those who are not chosen will not actually change.

As such, Goodin argues that the AAP should apply to all those who are potentially affected by the outcome of a decision. But the problem is that, for some decisions, adopting a potentially affected stance means incorporating vast numbers of actors into the decision, many of whom are unlikely to actually be affected by the outcome of the decision. In the example of the wind turbine, this would mean that, out of all the possible option sets for its location, all those who are potentially affected by the noise of the turbine should participate in the decision about where to site the turbine. However, this seems an overly demanding condition for participation, especially in light of the practical issues that are likely to arise in meeting this requirement. In response, David Miller argues that a group only has a justifiable claim for inclusion if their interests would be significantly affected by its decisions whichever way those decisions go (Miller 2009). Yet this only resolves the problem in some situations. There are still some decisions that potentially affect very large numbers of actors, each of whom are potentially affected in a significant way.

This leaves the AAP in a tricky bind. It is necessary to decide whether the AAP implies that all those who are *actually* affected, or all those who are *potentially* affected should participate in a decision, or if the AAP should be abandoned altogether when potentiality arises. The problem is that the AAP is silent on these issues and whilst these matters may not be irresolvable, they do involve some difficult decisions about how this principle can be achieved in practice.

A third problem with the AAP is much more of a concern for its overall plausibility as a principle of justice. This is that the AAP is counterintuitive in many everyday situations. For example, in the course of everyday life, many decisions have implications for other actors, yet people do not think that those who are affected by these decisions have any claim to inclusion in the way that they are made.[15] One might argue that, whilst merely being affected is not sufficient for inclusion, if you are significantly affected by a decision then you should have a say in how it is made. Yet on further inspection this doesn't seem a suitable approach either. Taking Robert Nozick's famous example, several suitors are significantly affected by the outcome of a bride's decision, yet people do not think that they should have a claim to participate in that decision (Nozick 1974, p. 269). In response to this, one might argue that the AAP is an appropriate principle in situations of public decision-making, rather than in private affairs. People often feel that some decisions should remain in the hands of certain actors, regardless of how these decisions affect

[15] Note that some authors argue against this view: Goodin 2007; Dryzek and Stevenson 2012a.

people. If a bride is choosing whom to marry, or if a university is deciding whom to offer a place to, then people feel that these actors should have a prerogative to make their own decision.

But there are also examples where agents are significantly affected by public decisions yet those actors do not have a right to participate in the decisions. Government decisions about immigration policy, the taxation of imported goods, or domestic environmental policies are all cases in which outside actors are significantly affected whilst lacking a legitimate claim to inclusion.

One reason that the AAP diverges from our everyday intuitions is that in many cases people can avoid being affected by the outcomes of a decision. Our intuitions about who should be included in a decision-making process often seem to depend on whether an actor can take action to avoid being affected by the outcome of a decision.[16] If a company purchases energy from a renewable energy supplier, then it is affected by the supplier's decision to raise the price of energy. But it seems strange to demand that the company should have a say in the supplier's decision as a matter of fairness. Likewise the supplier is affected if the company sources its energy from elsewhere, yet it doesn't seem a pressing matter of fairness. Regardless of what can be said about the practicality of implementing the AAP, it seems that it only appears to fit with our everyday intuitions in cases of unavoidability. This suggests that it's necessary to consider a different criterion for inclusion that takes this feature into account, rather than rely on affectedness itself.

These problems present the AAP with different challenges. The first two problems are not irresolvable in themselves. Whilst it's necessary to have a more nuanced understanding of what it means to be affected and what it means to participate in a decision, this is not to say that the AAP is inappropriate for procedural justice. Likewise, the issue of potentiality can be resolved through further refinement of what the AAP actually requires. In many cases we're likely to have some initial idea of who'll be affected by a decision, and those who are likely to be affected could be included in a decision, without having to include absolutely everyone for every single decision.

But the third problem is much more troublesome. It's necessary to find a more accurate account of what it means to be affected in a way that provides a legitimate claim to inclusion in a decision-making process, as well as a more thorough explanation of why affectedness might warrant inclusion in the first place. In what follows, I present two potential interpretations of affectedness that might fulfil this requirement: being subject to the law, and coercion. After considering each of these interpretations, I argue that the notion of coercion fits much more accurately with our everyday intuitions about who should participate in a decision.

[16]David Miller convincingly argues that the plausibility of the AAP as a principle of justice diminishes in such cases (Miller 2009, p. 218).

4.1.3 Subject to the Law

One form of affectedness that might fit better with our everyday intuitions is to be subject to an authority or power. The All Subjected Principle is a revised version of the AAP, which specifies that those who are subject to an authority should have some say in the way that its decisions are made.[17] There are different interpretations of what 'being subject to an authority' implies here. This can mean, for example, being subject to a law, or rule (what I call, the 'direct version of the All Subjected Principle').[18] But those who advocate the All Subjected Principle do not always refer to it in terms of subjection to the law. Nancy Fraser takes the view that an actor is subject to a governance arrangement if the arrangement subjects the actor to socioeconomic forces that are beyond his or her control (what I call, the 'indirect version of the All Subjected Principle'), rather than to the law per se (Fraser 2008, p. 96). This takes into account those who are effectively subject to the rules of an authority, whether or not they are actually subject to those rules in a formal sense.

In respect to the direct version, the All Subjected Principle seems quite intuitive here. There are many situations in which people feel that those who are subject to the law should have a say in the way that it is made. Within a domestic constituency, people typically feel that those who are subject to a law should have a say in the way that its made. There might be a good case for adopting the All Subjected Principle as a principle of procedural justice because participation legitimises the law to those who are subject to it. By participating in a decision, those who are subject to the law provide some form of consent (albeit tacit) to the law itself, giving the law legitimate authority, as well as providing those who are subject to the law a reason for complying with it.[19] The All Subjected Principle also fits with our intuitions about accountability. People typically feel that those who govern should be held to account by those who are governed. The direct version of the All Subjected Principle meets this requirement, by providing those who are governed by the law some control over those who make it.

But on further reflection, there are two problems with the direct version of the All Subjected Principle. First, being subject to the law requires a common set of rules, or a common authority that agents are subject to. That is, agents are only subject to the law when there is an established framework that subjects a unified community of actors to binding rules. Within a domestic state, there is a group of citizens with a common identity and a sovereign authority that can impose legal measures. But in many cases there is no such common community or sovereign power. It is well documented that there is no global sovereign power capable of

[17]Fraser 2008, p. 96; Karlsson 2008, p. 17, p. 80; see also: Goodin 2007, p. 42; Näsström 2011, p. 119. David Miller discusses this idea as an extension of the All Coerced Principle, which I discuss in the next section (Miller 2009, p. 222).

[18]For example: Karlsson 2008, p. 17.

[19]For accounts of legitimacy in international law, see: Bodansky 1999; Buchanan and Keohane 2006.

coordinating state action and enforcing compliance at the state level, nor is there any common community of citizens on whom the law can be placed (Miller 2010a). At the same time, it's often necessary to implement policies that govern behaviour at the global level, even if there is no common community on whom a legal framework can be imposed. This point does not mean that the direct version of the All Subjected Principle should be rejected, but rather that it should be supplemented in certain circumstances. That is to say, even if the All Subjected Principle is a sufficient principle of inclusion, it is not a necessary condition for inclusion here.

Second, there are at least some decisions that affect actors in significant ways without necessarily subjecting them to a law or rule. People often think that these actors have an entitlement to participate in some of the decisions that affect them, regardless of whether they are subject to the law. Multilateral institutions implement rules that have significant implications for many actors, few of whom are actually subject to these rules. If a state imposes a law on domestic manufacturers, then this might have significant implications for foreign suppliers. If these suppliers are sufficiently affected by this law, then they might feel that they are entitled to a say in the way that it is made, or at that our interpretation of the All Subjected Principle is insufficiently narrow here. It seems that direct version of the All Subjected Principle fails to capture all of our intuitions regarding who should participate in a decision-making process.

In response to these problems, one might adopt a broader notion of subjection along the lines that Fraser advocates. At the same time, adopting Fraser's indirect version of the All Subjected Principle is not without its problems. It opens the door to including a lot of actors who may only be subject to a governance structure in spurious, or tenuous ways. One might claim that foreign suppliers are owed some form of participation if they are sufficiently subject to a domestic policy. Yet this claim seems much less plausible for those who supply raw materials, transport needs, and accountancy services to these foreign suppliers also fall under this umbrella.

Adopting this notion of subjection also raises questions about why being 'subject' to an authority is important in the first place. If we care about the impacts of a decision or authority, then why are we concerned with who is subject to that authority, rather than who is actually affected by that decision, or burdened by its implications? Unless subjection means 'acting in our name and issuing laws in that capacity' (the direct version) then the All Subjected Principle is an inaccurate way of capturing what's really important. There are other notions of affectedness that are more appropriate for our intuitions about who should participate in a decision-making process.

On closer inspection then, the All Subjected Principle appears less convincing as an appropriate principle of procedural justice. It is too ambiguous and it is not clear what it is about being subject to a rule that warrants participation in a decision-making process. Whilst the All Subjected Principle does fit with our intuitions in certain situations, procedural justice needs a more refined notion of affectedness.

4.1.4 Coercion

If procedural justice requires a clearer definition of what it means to be affected in a way that requires participation in a decision, then coercion is a potential candidate. As David Miller points out, the affected interests principle is most plausible where people are unavoidably affected by certain decisions (Miller 2009, p. 217). Taking this into account, the All Coerced Principle states that all those who are coerced by the outcome of a decision should participate in the way that it's made.[20]

Before considering the relative merit of this approach, it's necessary to clarify what coercion means. Following Miller (who uses Grant Lamond's account of coercion as a starting point) a coercive act is one that meets three conditions: (i) it forces an actor to do something against their will, (ii) it subjects an actor to the will of another, and (iii) the coerced actor is unable to do otherwise (Lamond 2000, p. 43). Miller argues that what is normatively distinct about coercion is that it undermines the autonomy of the person who is coerced.[21] People feel that agents should be included in decision-making processes because autonomy and independence are valuable.

Following Joseph Raz's account of autonomy, a person is autonomous if they (i) have an adequate mental faculty, (ii) have an adequate range of options to choose from, and (iii) are not subject to the will of another (Raz 1988, p. 154). Autonomy is valuable here because it represents the ability to live an independent life, of one's own choosing. Coercive decisions undermine autonomy by preventing an actor from meeting either or both of the second and third requirements of autonomy. If a decision is coercive, then it undermines an agent's autonomy and it's necessary to rectify the situation in some way.

Typically, people suggest that there are three ways that this can be done. People could avoid taking the decision in the first place. Alternatively, people could provide some form of compensation to the coerced actor.[22] But it isn't always possible to avoid making a coercive decision, or to provide compensation in every case. So a third option is to justify our decision in some way. The All Coerced Principle is based on this third approach. If a decision impacts an agent's autonomy, then that agent is owed some special justification for the decision (Miller 2009, p. 219). This is not to say that people shouldn't take either of the other responses to coercion that I have also suggested here (avoiding the outcome or providing compensation). Infringing an agent's independence or autonomy is only permissible

[20]For discussion of this principle, see: Arrhenius 2005, p. 214; Abizadeh 2008; Miller 2009, p. 218.

[21]Terry Macdonald also argues that actors are entitled to participate in decisions that have an impact on their autonomous capacities (Macdonald 2008, p. 40).

[22]Robert Goodin suggests that people should provide compensation for any harm that they inflict upon those who are not part of a decision-making body (Goodin 2007, p. 68). Goodin's argument concerns harms rather than coercion, but the same principle applies here.

in specific circumstances, and people should either avoid coercive acts altogether or do their best to rectify them where this isn't possible. But sometimes it isn't possible to avoid making decisions that coerce people, nor is there a suitable way of providing compensation. In such cases, people owe the coerced actor some form of justification for their decision.

If our concerns about coercion are grounded in autonomy, then the form of justification that's required should reflect this fact. One way of respecting someone's autonomy is to gain his or her assent, and one way of doing this is to gain someone's assent to a decision by including him or her in a fair decision-making process. That is, an actor can come to accept the outcome of a decision if they are given a say in the way that the decisions is made. The upshot of this is that if an actor is coerced by a decision, then procedural justice requires that that actor have some say in the way that the decision is made. This gives us the All Coerced Principle**:** if a decision coerces an actor then that actor should have some say in the way that the decision is made

To illustrate this point, it's worth considering David Miller's response to Arash Abizadeh's argument for the democratic accountability of border controls.[23] Contrary to Abizadeh, Miller argues that border measures are not coercive because they leave potential immigrants with a suitable range of alternative options in order to achieve an autonomous life. If an individual wants to enter a particular state and they are prevented from doing so, then one might think that he is coerced by the decision to implement border controls because he is prevented from entering the country that he wants to live in. But this conflates the absence of one option with the absence of independence. As Miller rightly argues, the decision to close borders leaves a range of likewise options available to the individual, even if the exact option that he wanted is closed.

This example also highlights the important distinction between coercion and prevention (Miller 2009, p. 220). Prevention limits certain options whilst leaving a suitable range of alternative options open to an actor. This is insufficient to warrant participation because, according to our earlier account, it doesn't infringe autonomy. To be sure, prevention may require some other form of justification. But it's doesn't require that we give an actor a say in a decision because it doesn't affect autonomy in the way that coercion does. Coercion, on the other hand, limits the options available to an actor to the extent that they are forced to take an action that they would not have otherwise taken. Coercion affects our autonomy in a way that requires a say in how a decision is made. Prevention, on the other hand, doesn't require us to justify our decision to the same extent.

To illustrate this point, imagine that an institution imposes a mitigation policy that limits the demand of fossil fuels among its members. If this policy is large enough to affect the global demand of fossil fuels, then oil producers outside of the policy zone face a drop in demand for the goods that they produce. If an oil

[23]Miller 2010b; see also, Abizadeh 2008.

producer is still able to sell its products in other parts of the global market, then the policy is not coercive, because it leaves the producer with a likewise range of alternative options. Similarly, if the policy is sufficiently large to restrict the demand of fossil fuels in all global markets, but the oil producer has alternative means of maintaining its autonomy (for example, by switching its economy to tourism, or renewable energy production) then the policy is not coercive, because it leaves the oil producer with a suitable range of alternative options to lead an independent existence. But if the oil producer has no other means of generating income, then the institution's decision is coercive, because it affects the oil producer's autonomy.

Before moving on, it's worth saying something about potentiality. As I suggested earlier, including actors in a decision on the basis of the AAP is problematic, in part because, for some decisions, the decision-making process determines who is actually affected by the decision. Given that people should include those who are actually coerced by a decision, what should be done about those who are potentially coerced by a decision? The first thing to say is that there are different interpretations of what it means to be potentially coerced by a decision (just as there are different interpretations of what it means to be potentially affected). Potential coercion is important when a decision is actually made, rather than when something is merely under discussion. If a decision potentially coerces an actor, then people should listen to their views and interests so that they can make an informed and accurate decision. But this doesn't mean that these actors should be included in the decision-making process. Participation is only required because actors are actually coerced by decisions that infringe their autonomy. For this reason, it is necessary to listen to those who are potentially coerced by our decisions, rather than include them in the decision-making process.

4.1.5 Affected Interests

So far, I've argued that those in whose name a decision is made, and those who are affected by a decision should have some say in the way that a decision is made. Yet these principles provide an incomplete account of what's needed for fairness. Whilst the All Coerced Principle fits with many of our everyday intuitions, there are occasions when people owe an actor some justification for a decision, even if that actor's autonomy is not undermined. If a factory decides to adopt a more polluting production process then this may be very harmful for the inhabitants of a nearby town, who now have to take costly measures to clean up this pollution. But no one's independence or autonomy is undermined here. The residents of the town can continue to live independently, free from the will of other actors; they just face a cost on account of someone else's decision. All the same, given that these actors have been disadvantaged by the factory's decision, it seems unfair that they are excluded from the decision to change the production method. Procedural justice also requires that people listen to those whose interests are affected by their decisions. That is, those whose interests are potentially affected by a decision have the right to express their interests and views when that decision is made.

Someone's interests are affected by decisions that: prevent them taking options that they would otherwise take, leave them better or worse off, or impose negative externalities or other costs and burdens on them. Whilst those whose interests are potentially affected do not have a right to participate in these decisions (as they do for the All Coerced Principle), they still have a right to voice their views and concerns in the discussions, as well as a right to be heard.

The reasoning for this is as follows. Some decisions are legitimate, even if they disadvantage people. If people make decisions that negatively affect an agent's interests then procedural justice requires that they show that they have fully considered that agent's interests in their decision. This is also the case if these people could have made someone better off but chose not to. That is, fairness requires that people show these actors that they have not made the decision lightly, nor without due consideration of how those who are affected by our decision are burdened by it. People can only do this if they are aware of what these actor's interests actually are.

This alone doesn't require that decision-makers listen to those whose interests are affected by their decisions, or grant them a voice in a decision. It's possible to get an accurate account of their interests by listening to some other agent, or by making a judgement about how they are affected. It is the fact that decision-makers should treat those whose interests are affected with due respect that means that decision-makers should listen to them and grant them a voice in discussions. Treating an actor with the respect required by procedural justice means letting that actor represent his or her own interests and views regarding the decision, rather than getting this information from elsewhere. This also has an epistemic benefit, as listening to how people are affected allows us to make better decisions.

This principle extends to those whose interests are *potentially* affected by a decision. This is because, in some situations, people don't know whose interests are affected by their decisions until they know what the outcome of that decision is. Given that people should listen to those whose interests are actually affected by their decisions, this means that they should listen to all those whose interests are potentially affected. This also extends to those whose interests aren't necessarily affected by a decision, but could have been.

This provides a more nuanced account of who should participate in a decision-making process. This differs from the AAP because it gives stronger requirements for participation in a decision-making process. It differs from the All Coerced Principle in at least two ways. First, it suggests that there are more ways of being affected that require justification than coercion alone. Second, it suggests that justification does not just relate to formal power in a decision-making process, but also concerns listening to the voices and opinions of people, and taking their interests into account.

4.2 Affectedness, Coercion, and the UNFCCC

Having thought about what circumstances require the participation of certain actors in order for decisions to be fair, it's now time to turn our attention to what this might mean for UNFCCC. Given how decisions currently take place, procedural justice should act as a guiding principle for thinking about how policy-making could be improved in terms of fairness.

The UNFCCC makes decisions that affect the global atmospheric concentration of greenhouse gases. These decisions relate to global temperature ranges, or emissions targets that it should achieve, and they have significant implications for actors on a global scale. There's already a growing number of increasingly in-depth empirical accounts of the different ways in which these decisions already affect actors and how they will continue to do so in the future.[24] The physical effects of climate change include sea level rise, ocean acidification, changing weather patterns and increases in the geographical range of infectious diseases. As such, the outcomes of decisions made in the UNFCCC may bring about severe and irreversible changes to the climate system, which could have extreme consequences for fundamental human interests on a global scale.[25]

The outcomes of the UNFCCC aren't limited to its effect on the climate. The methods that the UNFCCC employs to achieve emissions targets also have significant implications for actors. These measures include: regulations and standards, taxes and charges, or tradeable permits.[26] The UNFCCC may impose measures on its member states, which in turn impose measures on their domestic populations. These sorts outcomes are likely to have wide reaching spill over effects that will have serious implications for people's wellbeing on local and global levels (AR5 WG3, p. 65). Reducing emissions in one policy area is also likely to have significant implications for those actors in other policy areas. Emissions policies are enacted in the context of global trade, and any reduction in the consumption of goods that are produced using carbon intensive processes in one policy area reduces the demand for these goods elsewhere in the world. The negative implications of domestic mitigation policies for those who are dependent on the production of such fossil fuels are well documented.[27] But institutional decisions about mitigation policies are likely to have different implications for different actors on a global scale.

But not all of the decisions made in the UNFCCC will have global implications. Decisions concerning adaptation in particular are likely to have significant implications for some actors, even if this isn't on a global scale. These measures may include infrastructure developments designed to help states deal with the changing climate, such as flood defence systems or changes to agricultural policies. If a small

[24] See the contributions of Working Group II to the IPCC Assessment Reports.

[25] Caney 2009; OHCHR 2009; Bodansky 2012.

[26] For definitions of selected mitigation policy instruments, see: Gupta et al. 2007, p. 70.

[27] Brandi 2010; Helm 2012, p. 12.

number of states cooperate to build a flood defence system, then this project may have severe implications both for actors within those states, as well as those who are outside of the institution. This might involve for example, the forced relocation of some actors, changes in the physical environment, or other significant costs.[28]

The point of this is that the UNFCCC will have to make a large number of decisions in relation to climate change. These decisions will have implications on a global scale, often in extremely profound ways. Some of these decisions will have outcomes that will ultimately affect actors in such a way that they are coercive. If the UNFCCC collectively adopts a global temperature threshold that creates sea level rise to the extent that entire nations become uninhabitable then it seems wholly plausible that the citizens of that state are coerced by its decision (following the definition of coercion outlined earlier).

One might claim that the UNFCCC should avoid making decisions that coerce people altogether. But the problem with this is that some coercive decisions are unavoidable for achieving certain important ends. Avoiding something like dangerous climate change inevitably involves making decisions that impose coercive externalities on those around the world. Climate change is simply too far-reaching, and too embedded in our society to avoid this. The absence of a coerced actor from a decision shouldn't stop the UNFCCC from making decisions. This is because addressing climate change requires making some decisions that are unavoidable. Whilst fairness sometimes requires that people are included in a decision, their absence shouldn't stop us from taking action on important matters. Sometimes, the stakes are too high to limit our decisions to those that every actor chooses to participate in. Given the high stakes involved with climate change, fair participation means that the UNFCCC should attempt to justify its decisions to these actors, but it doesn't mean coerced actors can hold progress up.

The account that I give here is by no means exhaustive, and only serves as a rough guide for how people should start to think about the decisions of the UNFCCC. But it shows that there are at least some cases where people have very strong claims to a say in the way that the decisions of the UNFCCC are made. In some cases, this claim might be strong enough to mean that actors should have some sort of formal say in the decision by virtue of how much the decision affects them. Given that climate change is likely to affect large numbers of actors in extreme ways, procedural fairness requires that the UNFCCC should have a high level of representation in its decisions.

So its necessary to consider about how we might interpret this claim to a say in a decision in a multilateral context. Given what's been said so far about the procedural problems associated with decisions amongst large groups, it doesn't seem worthwhile spending too much time considering whether all those affected

[28]For discussions concerning the impacts of adaptation measures see: Thomas and Twyman 2005; Huq and Khan 2006.

should have a direct say in the UNFCCC's decisions.[29] What seems much more reasonable is to suggest that those who should have a say in the way that a decision is made should have their views and interests represented in that decision. In this respect, those who should have a say in decisions at a multilateral level can have a say in these decisions through some form of representation; whether this is through just the representation of their views in a debate, or a more formal stake in decisions through a vote.

Support for the nation that individuals can be represented in multilateral decision through the presence of views and discourses can found in the literature on 'discursive representation'. This includes the studies by Dryzek and Niemeyer, which Dryzek and Stevenson have subsequently considered in the context topic of climate change.[30] This literature suggests that it's possible to replicate the actual participation of actors in deliberative decisions by ensuring that all relevant discourses are present in discussions, where a discourse is a set of information that captures the diversity of values, interests and needs of all affected actors.[31] Advocates of discursive representation argue that incorporating these discourses allows us to reach a second best alternative in multilateral negotiations when giving actors an actual voice in these discussions is very difficult.

Yet even if we adopt this notion of fair participation, it's clear that the COP negotiations fall far short of representing all those who should have a say in its decisions. Currently, procedural rights to speak and vote in the decisions of the COP remain firmly in the hands of the UNFCCC's member states. On the one hand, this limits the representation of decisions if you're not represented by a state. On the other hand, this also falls short of an ideal notion of fair representation. The accurate and fair representation of all views in multilateral decisions is elusive. Even the most democratic states have poor levels of representation at the multilateral level. Even those states that are democratically accountable might not sufficiently represent the interests of those that they claim to. Many actors have little democratic accountability at the multilateral level. NSAs that represent large numbers of actors do so without any authority or credibility. If the UNFCCC makes decisions that affect actors in ways that require their participation in its decision-making processes, and if the UNFCCC falls down in this respect, then it's worth thinking about how to improve participation in COP negotiations. In the following section I discuss several measures that might improve procedural fairness in the UNFCCC.

[29]Although, some authors have considered whether individuals on a global scale should have a direct vote in multilateral decisions.

[30]See: Dryzek and Niemeyer 2006; Dryzek and Stevenson 2011, p. 1868, 2012a, b.

[31]Dryzek and Stevenson 2011, p. 1868.

4.3 Fair Participation in the UNFCCC

Foremost, the UNFCCC should further encourage the involvement of NSAs in its decisions as agents that represent actors that are otherwise missing from the debate. This means that NSAs, such civil society groups and NGOs, should play a large role in influencing and shaping the debates of the UNFCCC. This is because climate change is likely to have profound effects for people on a global scale and those whose interests are affected by a decision should have their interests advocated during the decision-making process. By encouraging states to engage with NSAs in negotiations more generally, the UNFCCC can ensure that a wide range of views and interests are heard in the debate. There is a lot of support for the view that NSAs can play an important role in this regard.[32]

This doesn't, however, mean that the UNFCCC should encourage the participation of unlimited numbers of actors in its negotiation forum, nor that NSAs should be granted the right to make interventions in negotiations without invitation. As I've said throughout this chapter, COP negotiations are already heavily strained from the pressure of large numbers of actors claiming to have a stake in the way that its decisions are made. Further burdening COP negotiations by increasing the representation of NSAs would be a mistake. Moreover, permitting more actors to make interventions in these debates would place a prohibitive burden on negotiations that already run wildly over-schedule. Formal interventions are neither necessary, nor practical measures to achieve the added value of NSAs in this context. What's more important is that these actors have a say in the way that debates are shaped at some point along the line. This doesn't have to happen in the COP forum itself. COP negotiations represent a culmination of work that takes place over many months, or even years, as participants gather evidence and form positions ahead of the formal negotiation sessions. In this respect, it's more important that NSAs have the chance to be heard at an earlier stage in the shaping of debates, by offering viewpoints and providing a voice in the debate that would otherwise be absent.

An example of how this could be done comes from civil society engagement in the European Union. Here, NSAs representing civil society have achieved significant representation in EU institutions though the European Economic and Social Committee, a diverse collection of civil society representatives that advises the decision-making bodies of the European Union. This gives civil society actors the opportunity to present views, discuss decisions and form part of the dialogue without unduly holding up decisions.

But this isn't the only role that NSAs should play. As I argued earlier, some actors should have a right to much more of a say in the way that decisions are made than just having their voice heard. This includes those who are coerced by a decision, or in whose name a decision is made. There may be good grounds for giving NSAs a

[32]Karen Bäckstrand argues that NSAs have extremely important roles for multilateral institutions even if they lack formal power in these contexts (Bäckstrand 2006, p. 484.

formal say in decisions if they act on behalf of these groups. Some might claim that this is too idealistic, and that it's simply unfeasible to award those who can't take on obligations a formal say in this context. But there are other ways of offering this sort of power.

But there's a further reason for being wary about giving NSAs a say in the UNFCCC, which is arises from a concern about the democratic legitimacy. Just as we should be wary that some states are sometimes unaccountable to their citizens, we might also be concerned about the participation of some NSAs and civil society groups on the same grounds. Many actors participate in the UNFCCC COP, as well as other multilateral forums, on the basis that they represent groups that are otherwise forgotten voices in debates that profoundly affect them. Yet many of these actors participate in these decisions with a minimal degree of democratic accountability. Given that we want these actors to participate in decisions by virtue of their representation, this is deeply concerning for thinking about procedural fairness. This gives a second policy recommendation for fair participation, which is that the UNFCCC should scrutinise the democratic legitimacy of both state and non-state actors in its decisions.

The fact that many NSAs participate in multilateral negotiations without any direct accountability is a cause for concern here, and the UNFCCC should try to reduce inaccurate representation where possible. For example, accreditation requirements could be imposed on representative agents.[33] These agents could be expected to show that they are acting on behalf of the individuals that they claim to wherever it is possible. This might include, for instance, evidence of public support, or popular approval by those who are represented. Biermann et al. support this view, arguing that, whilst giving greater rights to NSAs in multilateral governance institutions is desirable, the UNFCCC should also try to establish more effective accountability mechanisms between these agents and the constituencies that they represent (Biermann et al. 2012).[34] At the same time, it should be recognised that democratic accountability isn't always possible. Whilst stringent rules governing the accountability of representative agents are desirable, the fact that some groups are disparate or difficult to identify means that the UNFCCC shouldn't demand strong democratic standards in every instance.

But this isn't the only way to improve the democratic legitimacy of representative actors participating in the UNFCCC. In many cases it won't be possible or feasible to provide high levels of credible accountability. Many states that participate in negotiations aren't democratic, and it seems overly optimistic to think that efforts at the multilateral level will have any success in changing this. In other cases, NSAs might be incapable of providing the level of democratic accountability that we'd ideally expect of an agent acting on behalf of individuals in COP negotiations. So

[33]For a discussion of the participation and accreditiation of NGOs in multilateral agreements, see: Oberthür et al. 2002, p. 130.

[34]For more on this point, see: Scholte 2002; Moore 2006, p. 34.

it's also worth thinking about how else the decisions of the COP can be made more representative and accountable to those affected by its decisions.

This leads to a third policy recommendation of the procedural rules of the UNFCCC, which is that COP negotiations should be made more accessible to individuals globally. The reasoning for this is that, for the moment, democratic participation in multilateral decisions is highly challenging. One way of making these decisions more representative, however, is to increase the understanding of what's being decided in these decisions globally. By disseminating information about its decisions and how these will ultimately affect people, the UNFCCC can take a step towards improving the overall representation and accountability of its decisions by empowering individuals globally to challenge those who act on their behalf in decisions.

There are a number of measures that the UNFCCC could adopt to improve this. For example, it could broaden its own communication and information dissemination about its decisions. It could engage more with civil society actors at the domestic level to encourage that more views to be brought forward from the bottom level up. These represent initial ideas about what could be done in order to improve representation in the decisions of the UNFCCC. But the important point here is that, whilst representation in global politics will always operate under the constraints of what's feasible and practical, there are at least some measures that can taken to improve the representation deficit within the UNFCCC.

4.4 Conclusion

The procedural deadlock that persists in the UNFCCC is deeply concerning. The fact that we need to achieve effective action on climate change quickly means that it's worthwhile considering ways of reducing the problems of trying to reach agreement amongst large numbers of actors. At the same time, the UNFCCC should remain representative and fair. Often these goals are presented as mutually exclusive: either we achieve effective action at the cost of fair representation, or we continue to face stalemate in inclusive negotiation processes. In this chapter, I've proposed a way of taking both of these matters into account.

References

Abizadeh, A. 2008. Democratic theory and border coercion: No right to unilaterally control your own borders. *Political Theory* 38: 111–120.

Agné, H. 2006. A dogma of democratic theory and globalization: Why politics need not include everyone it affects. *European Journal of International Relations* 12(3): 433–458.

Archibugi, D., D. Held, et al. 1998. *Re-imagining political community: Studies in cosmopolitan democracy*. Cambridge: Polity Press.

Arrhenius, G. 2005. The boundary problem in democratic theory. In *Democracy unbound*, ed. F. Tersman. Stockholm: Stockholm University.

Bäckstrand, K. 2006. Democratising global governance? Stakeholder democracy after the world summit on sustainable development. *European Journal of International Relations* 12(4): 467–498.

Bäckstrand, K. 2010a. Democratizing global governance of climate change after Copenhagen. In *Oxford handbook on climate change and society*, ed. J. Dryzek, R.B. Norgaard, and D. Schlosberg. Oxford: Oxford University Press.

Bäckstrand, K. 2010b. *Environmental politics and deliberative democracy: Examining the promise of new modes of governance*. Cheltenham: Edward Elgar.

Banuri, T., K. Goran-Maler, et al. 1995. Equity and social considerations. In *Economic and social dimensions of climate change*, Contribution of working group III to the second assessment report of the Intergovernmental Panel on Climate Change, ed. J.P. Bruce, L. Hoesung, and E. Haites. Cambridge: Cambridge University Press.

Beerbohm, E. 2012. *In our name: The ethics of democracy*. Princeton/Oxford: Princeton University Press.

Biermann, F., K. Abbott, et al. 2012. Navigating the anthropocene: Improving earth system governance. *Science* 16.335(6074): 1306–1307.

Bodansky, D. 1999. Legitimacy of international governance: A coming challenge for international environmental law? *The American Journal of International Law* 93(3): 596–624.

Bodansky, D. 2012. *The durban platform: Issues and options for a 2015 agreement*. Centre for Climate and Energy Solutions: http://www.c2es.org/docUploads/durban-platform-issues-and-options.pdf

Bodansky, D., and L. Rajamani. 2013. Evolution and governance architecture. In *International relations and global climate change*, ed. D. Sprinz and U. Luterbacher. Cambridge, MA/London:MIT Press.

Brandi, C. 2010. *International trade and climate change: Border adjustment measures and developing countries*. Bonn: Deutsches Institut für Entwicklungspolitik/German Development Institute.

Buchanan, A.E., and R.O. Keohane. 2006. The legitimacy of global governance institutions. *Ethics and International Affairs* 20(4): 412.

Bulkeley, H., and P. Newell. 2010. *Governing climate change*. Abingdon/New York: Routledge.

Caney, S. 2009. Climate change, human rights and moral thresholds. In *Human rights and climate change*, ed. S. Humphreys and M. Robinson. Cambridge: Cambridge University Press.

Dahl, R.A. 1975. Procedural democracy. In *Philosophy, politics and society*, ed. P. Laslett and J. Fishkin. Oxford: Blackwell.

Dimitrov, R.S. 2010. Inside Copenhagen: The state of climate governance. *Global Environmental Politics* 10(2): 18–24.

Dombrowski, K. 2010. Filling the gap? An analysis of non-governmental organizations responses to participation and representation deficits in global climate governance. *International Environmental Agreements: Politics, Law and Economics* 10(4): 397–416.

Dryzek, J., and S.J. Niemeyer. 2006. Reconciling pluralism and consensus as political ideals. *American Journal of Political Science* 50(3): 634–649.

Dryzek, J., and H. Stevenson. 2011. Global democracy and earth system governance. *Ecological Economics* 70(11): 1865–1874.

Dryzek, J., and H. Stevenson. 2012a. The discursive democratization of global climate governance. *Environmental Politics* 21(2): 189–210.

Dryzek, J., and H. Stevenson. 2012b. Legitimacy of multilateral climate governance: A deliberative democratic approach. *Critical Policy Studies* 6(1): 1–18.

Falk, R., and R. Strauss. 2000. On the creation of a global people's assembly: Legitimacy and the power of popular sovereignty. *Stanford Journal of International Law* 36: 191–220.

Fraser, N. 2008. *Scales of justice: Reimagining political space in a globalizing world*. Cambridge: Polity.

Goodin, R.E. 2007. Enfranchising all affected interests and its alternatives. *Philosophy & Public Affairs* 35(1): 40–68.
Gupta, S., D.A. Tirpak, et al. 2007. Policies, Instruments and Co-operative Arrangements. In *Climate change 2007: Mitigation. Contribution of working group III to the fourth assessment report of the Intergovernmental Panel on Climate Change*, ed. B. Metz, O.R. Davidson, P.R. Bosch, R. Dave, and L.A. Meyer. Cambridge: Cambridge University Press.
Haas, P.M. 2008. Climate change governance after Bali. *Global Environmental Politics* 8(3): 1–7.
Held, D. 2004. *Global covernant*. Cambridge: Polity Press.
Helm, D. 2012. The Kyoto approach has failed. *Nature* 491: 663–665.
Huq, S., and M.R. Khan. 2006. Equity in National Adaptation Programs of Action (NAPAs): The case of Bangladesh. In *Fairness in Adaptation to Climate Change*, ed. N.W. Adger, J. Paavola, S. Huq and M.J. Mace. Cambridge, MA/London: MIT Press.
Jagers, S.C., and J. Stripple. 2003. Climate governance beyond the state. *Global Governance* 9(3): 385–99.
Karlsson, J. 2006. Affected and subjected-the all-affected prinicple in transnational democratic theory. *Social Science Research Center Berlin*. Discussion Paper SP IV 2006–304.
Karlsson, J. 2008. Democrats without borders: A critique of transnational democracy. In *Gothenburg studies in politics*, ed. Bo Rothstein. Gothenburg: Department of Political Science, University of Gothenburg.
Kravchenko, S. 2010. Procedural rights as a crucial tool to combat climate change. *Georgia Journal of International and Comparative Law* 38(3): 613–48.
Lamond, G. 2000. The coerciveness of law. *Oxford Journal of Legal Studies* 20(1): 39–62.
Lawrence, P. 2014. *Justice for future generations*. Cheltenham: Edward Elgar
Lijphart, A. 1984. *Democracies: Patterns of majoritarian and consensus government in twenty-one countries*. New Haven/London: Yale University Press.
Macdonald, T. 2008. *Global stakeholder democracy: Power and representation beyond liberal states*. Oxford/New York: Oxford University Press.
Miller, D. 2009. Democracy's domain. *Philosophy & Public Affairs* 37(3): 201–228.
Miller, D. 2010a. Against global democracy. In *After the nation: Critical reflections on post-nationalism*, ed. K. Breen and S. O'Neil. Basingstoke: Palgrave Macmillan.
Miller, D. 2010b. Why immigration controls are not coercive: A reply to Arash Abizadeh. *Political Theory* 38(1): 111–120.
Moore, M. 2006. Globalization and democratization: Institutional design for global institutions. *Journal of Social Philosophy* 37(1): 21–43.
Näsström, S. 2011. The challenge of the all-affected principle. *Political Studies* 59: 116–134.
Nozick, R. 1974. *Anarchy, state and utopia*. New York: Basic Books.
Oberthür, S., et al. 2002. *Participation of non-governmental organisations in international environmental governance: Legal basis and practical experience*. Berlin: Ecologic.
OHCHR. 2009. Report of the office of the United Nations high commissioner for human rights on the relationship between climate change and human rights. In *The office of the United Nations high commissioner for human rights and the office of the high commissioner and the secretary-general*. http://www.ohchr.org/Documents/Press/AnalyticalStudy.pdf.
Palerm, J.R. 1999. Public participation in environmental decision-making: Examining the Aarhus convention. *Journal of Environmental Assesment and Policy Management* 1(2): 229–244.
Rapp, T., C. Schwägerl, et al. 2010. The Copenhagen protocol: How China and India sabotaged the UN climate summit. *Der Spiegel*, 5th May 2010.
Raz, J. 1988. *The morality of freedom*. Oxford: Oxford University Press.
Scholte, J.A. 2002. Civil society and democracy in global governance. *Global Governance* 8(3): 281–304.
Shelton, D. 2007. Equity. In *Oxford handbook of international environmental law*, ed. D. Bodansky, J. Brunnée, and E. Hey. Oxford: Oxford University Press.
Thomas, D.S.G., and C. Twyman. 2005. Equity and justice in climate change adaptation amongst natural-resource-dependent societies. *Global Environmental Change* 15: 115–124.

Toth, F.L., M.J. Mwandosya, et al. 2001. *Decision making frameworks*, Climate change 2001: Mitigation. Contribution of working group III to the third assessment report of the Intergovernmental Panel on Climate Change. Cambridge: Cambridge University Press.

UNCED. 1992. United Nations conference on environment and development. Agenda 21. (The Rio Declaration). In *U.N. conference on environment and development*, Rio de Janeiro, Brazil, June 3–14, 1992.

UNECE. 1998. *United Nations economic commission for Europe*. Convention on access to information, public participation in decision-making and access to justice in environmental matters, June 25, 1998 (Aarhus Convention). Geneva: UNECE.

UNFCCC. 1992. *United Nations framework convention on climate change*. Convention Text.

Victor, D. 2006. Toward effective international cooperation on climate change: Numbers, interests and institutions. *Global Environmental Politics* 6(3): 90–103.

Vidal, J. 2009. *Secrecy prevails at Bangkok climate talks*. London: The Guardian.

Whelan, F. 1983. Prologue: Democratic theory and the boundary problem. In *Nomos XXV: Liberal democracy*, ed. J.R. Pennock and J.W. Chapman. New York/London: New York University Press.

Chapter 5
Political Equality: Levelling the Playing Field

5.1 Introduction

In Chap. 3, I argued that fair procedures have intrinsic value. I also proposed that fair procedures could lead us to find a mutually acceptable agreement when there is reasonable disagreement about the ends that the UNFCCC should achieve. In Chap. 4, I identified who should participate in the decision-making processes of the UNFCCC. Now it's necessary to say something about what fair procedures actually are. In order to do this, it's necessary to develop a more formal notion of procedural justice. Having done this, it's then possible to think about what procedural justice implies for *how* actors participate in the decisions of the UNFCCC.

To this end, this chapter has two aims. First, I explore the intrinsic value of procedural justice. In order to do this, I argue that there is a deep connection between procedural justice and democracy and that the same normative ideals form the basis of these two concepts. More specifically, I argue that fair procedures should be thought of as those that (i) allow decision-makers to participate autonomously, (ii) that let decision-makers advocate their interests equally in decisions, and that (iii) encourage decision-makers to justify their decisions to one another. Many theorists invoke these principles during their accounts of democratic theory and this chapter draws on these existing arguments in order to consider which values procedural justice is based on.

Having provided the normative basis for procedural justice, my second aim in this chapter is to discuss what this means for equality in the decision-making processes of the UNFCCC. Many people hold that fair procedures are those that invoke some notion of political equality (Beitz 1989; Verba 2003). This is something that's also frequently brought up in discussions of the fairness of COP negotiations. Several reports have expressed deep concerns about the ability of developing countries to participate in decisions throughout the history of the UNFCCC, suggesting that decisions are unfair either because the rules of the game are biased, or because there

L. Tomlinson, *Procedural Justice in the United Nations Framework Convention on Climate Change*, DOI 10.1007/978-3-319-17184-5_5

are significant resources asymmetries between actors.[1] These asymmetries give some give some decision-makers a significant advantage in being able to determine the outcome of a decision and lead to some of the most frequent criticism of the procedural fairness of the UNFCCC. This raises questions about what principles should govern decision-making in the UNFCCC, and what resources decision-makers should have access to.

My aim in this chapter is to answer these questions. I do this by developing a normative account of procedural justice and then to drawing on this to examine what role political equality should play in the decision-making processes of the UNFCCC, what form it should take, and what it entails in practice. I analyse these normative arguments in the context of the procedural difficulties that decision-makers actually face in the UNFCCC.

In doing so, I claim that political equality consists of two principles:

(1) Equal status among decision-makers
(2) Equal opportunity to influence decisions

Having defended this, I then make some substantive suggestions for how the UNFCCC can achieve these principles in practice, and outline several policy measures that the UNFCCC can employ to meet the principles of political equality requires described above. This includes the following recommendations for procedural reform:

(1) The COP Chair and the UNFCCC Secretariat should play lead roles in ensuring that state delegations are treated with equal respect and face a level playing field
(2) The UNFCCC Rules of Procedure should promote the role of Non-State Actors and domestic institutions in assisting state delegations in negotiations

5.2 Democracy and the Grounds for Procedural Justice

One of the aims of this chapter is to determine what procedural justice requires for decisions. This means thinking about the normative ideals that ground fair procedures. Some people suggest that this is quite straightforward. For example, Ben Saunders suggests that procedural justice is about giving each actor an equal say in a decision (provided that each actor has a roughly equal stake in the decision (Saunders 2010)). Similarly, when discussing political equality, Sydney Verba argues that fairness means that each should have an equal say in a decision-making process (Verba 2003), and Thomas Christiano argues that fairness requires that each citizen has an equal vote in majority voting processes (Christiano 1996, p. 3).

[1]See, for example: IPCC Special Report on Participation of Developing Countries (IPCC 1990; Chasek and Rajamani 2003; UNfairplay 2011).

But there's obviously more to fair procedures than having an equal say in a decision-making process. If we only cared about equality then we could just as well make a decision through a lottery or coin toss. Instead, many suggest fair procedures are those that also include a commitment to justification and autonomy (Gould 1988; Dryzek 2000, 2010). People usually think that some epistemic value is also important. Therefore, before making any claims about what procedural justice requires, it's necessary to say more about what actually grounds procedural justice.

If it's necessary to think about the normative grounds for procedural justice then one way to start is by considering some of the arguments given for democracy. In what follows, I show that democracy provides a useful starting point for thinking about procedural justice. I then consider some of the most prominent arguments for democracy, including arguments based on autonomy, justification, and equality. Having done this, I then outline the values that ground procedural justice.

There are already a lot of arguments about the underlying values that ground democracy and many of these consider how decision-making processes can treat people fairly.[2] If some of these arguments overlap with those for procedural fairness, then this should prove a helpful starting point for thinking about what grounds procedural justice. There are several reasons for thinking that the arguments that ground democracy overlap with those for procedural justice.

Most accounts of democracy start from the premise that it is necessary to make collective decisions among people who disagree. Societies are typically very diverse, and it's often necessary to make collective decisions among people who have conflicting views and opinions. Many accounts of democracy claim that democratic processes are those that are made collectively and that treat each actor fairly (Dahl 1956; Saunders 2010; Estlund 2008, p. 81). In this respect, these accounts share the same goals that fair procedures should achieve in the UNFCCC. In Chap. 3, I argued that procedural justice can help actors to reach a mutually acceptable outcome when there is reasonable disagreement. I argued that if there is significant initial disagreement between people then it's unlikely that they will be able to reach a consensus through reasoned discussion. Given that many accounts of democracy start from the premise that there is significant disagreement in society, this provides a useful place for thinking about what grounds procedural justice.

But it's not just that the UNFCCC needs procedures to enable those who disagree to make collective decisions; the UNFCCC needs procedures that each can perceive as sufficiently fair so that its outcomes are universally acceptable. In this respect, democratic procedures share a second motivation with procedural justice. For many democratic theorists, an important part of democratic decisions is that they bring about outcomes that everyone accepts and has a moral duty to comply with. Part of reaching an outcome that is morally binding is about developing a democratic process that sufficiently respects each actor so that each has a moral duty to comply with the outcome of that process. Given that our aim is to reach a procedure that

[2]Thorough accounts of the grounds of democracy given by: Kelsen 1955; Christiano 1990, 2008; Estlund 2008, p. 70.

each considers sufficiently fair so that its outcome is acceptable to all, arguments for democracy are likely to provide some ideas for the underlying values of procedural justice.

A further reason for looking at democracy is that some accounts of what grounds democracy appeal to the intrinsic value of procedures in reaching an outcome that all can accept as binding. That is, not only do these accounts of democracy share the ends that procedural justice should achieve, but they also do so by developing sufficiently fair procedures. Given that the UNFCCC needs procedures that are sufficiently fair that each decision-maker can accept the outcome, even if they disagree with the correctness of the outcome itself, democracy should provide one way of starting to think about what fair procedures should look like.

For these reasons, it is helpful to examine the normative grounds of democracy to gain an understanding of the value and nature of procedural justice. There are at least three prominent arguments for democracy that are worthwhile exploring as possible grounds for procedural justice: autonomy, equality, and justification.

Some argue that democracy is important on the basis of autonomy.[3] People generally think that autonomy is a fundamentally important feature of our lives, and participating in decisions allows people to have some sort of input over important decisions the affect their lives. Many democratic decisions have profound impacts on people's lives. By taking part in these decisions, people can act as authors of the rules that govern them. The important point is that people are able to maintain some degree of self-government, even in a society where it's necessary to make decisions that restrict people's autonomy in profound ways. In Chap. 4, I emphasised this point when I argued for the importance of participation in a decision-making process as a matter of autonomy. I argued that people have a right to representation in decision-making processes that coerce them, partly because coercion is damaging for autonomy and participation partially remedies this damage. This also fits with our thoughts about what's important in a democratic process. People usually think that having some sort of input into or control over a decision is important. That is, it's important that people actually vote in a decision, rather than just having their preferences collected and aggregated, even if they ultimately arrive at the same outcome.[4]

An alternative justification for democracy concerns equality. Egalitarian arguments often start by considering how people should address the problem of disagreement in society. Thomas Christiano advocates such an approach that focuses on giving each actor's interests equal consideration (Christiano 2008). Christiano argues that each individual has a fundamental interest in being treated as an equal in society. For Christiano, democracy is desirable because it is the only way of achieving this end when there is significant disagreement about justice in

[3]For example: Gould 1988. Jean-Jacque Rousseau's account of democracy is also based on autonomy (Rousseau 1762).

[4]David Estlund also makes this point, Estlund 2008, p. 76.

society (Christiano 1990, 2008).[5] Each individual's interests are intertwined with fundamental features of society and there are inevitable conflicts between different people's interests. Given this, one way of treating people's interests equally is to give each person an equal say in democratic decisions. Democracy does this by giving each individual the equal ability to advance his or her concerns in decisions about the organisation of society. According to Christiano, this leads us to a democratic process with features such as equal voting power, equality of opportunity to run for office, and equal opportunities to participate in democratic decisions.

A further argument for democracy is based on the importance of justifiability. This is basis for some arguments for deliberative democracy.[6] The idea behind deliberative democracy is that outcomes are legitimate to the extent that they are the result of free and reasoned deliberation among equal actors. The process of deliberation involves arriving at a consensus through the exchange of mutually acceptable reasons. Each decision-maker puts forward reasons that others can come to accept or, to put it another way, reasons that are justifiable to others. According to Gutmann and Thompson, when there is disagreement and reasonable pluralism, citizens owe each other justifications for the laws that they impose on one another as a basic moral ideal. In Gutmann and Thompson's terms, this notion of 'reciprocity' allows us to arrive at an outcome in a way that treats each actor with sufficient respect so that each can accept the outcome as binding (Gutmann and Thompson 2000).[7] The key point is that deliberative decisions produce legitimate outcomes, in the sense that these outcomes can be justified to others. Justification seems a promising ideal that democratic processes should aspire to. Because the aim of a deliberative process is to arrive at an outcome that each actor is willing to support, it shares some of our motives for designing a fair decision-making process.

There are of course many other reasons for thinking that democratic processes are important. In particular, some people claim that democracy is important for instrumental reasons, such as the desirable outcomes and externalities that it produces. But the arguments presented here represent the most prominent intrinsic arguments for democracy. They also seem to align with much of the thinking around what's important for democracy in the negotiations of the UNFCCC. As I argued in Chap. 4, actors attribute a great deal of value to being able to participate in COP negotiations as independent agents. These are matched by claims around equality and justification. Given that ideals of democracy can provide one way of thinking about the ideals that ground procedural justice, these arguments can help us think about the normative ideals that ground procedural justice. This gives us a platform

[5] See also, Christiano 1996, 2004.

[6] For more on deliberative democracy, see: Cohen 1989, 1997; Bohman 1997; Mansbridge et al. 2010.

[7] Joshua Cohen refers to this concept as deliberative inclusion (Cohen 1989, 1997). See also: Rawls 1993, p. 137.

for thinking about what fair procedures look like. In what follows, I claim that fair procedures are those that respect these values, as well as some other normative values.

The UNFCCC needs procedures that are sufficiently fair so that its outcomes are universally acceptable. Christiano's argument for equality provides one way of thinking about how we can achieve this goal. Christiano's argument focuses on the need to respect interests when it is necessary to make decisions about the arrangement of society, and when people disagree about how this should be done. Given that it allows us to reach an agreement that treats each decision-maker fairly, this seems an appropriate way for thinking about fair procedures in the UNFCCC. But this isn't the only thing that's important here. As I suggested earlier, there is also something important about the fact that decision-makers actually play a role in reaching these decisions. This suggests that fair procedures should also pay some respect to autonomy in the UNFCCC. It isn't enough that decision-makers' interests are advanced equally in a decision; the decision-makers themselves must play some role in the way that this happens.

Further to equality and autonomy, fair procedures are those in which decision-makers justify decisions to one another. In order to see this, it's worth recalling the argument from Chap. 3, where I suggested that fair procedures are those in which reasonable actors respect each other's claims. In this respect, it is not enough that decision-makers put forward claims in a decision-making process; decision-makers should justify their claims to one another in terms of reasonableness. This means that fair procedures are those in which decision-makers justify their decisions to one another. Justification implies that decision-makers respect one another and offer reasons that others can accept as reasonable, even if they disagree with the correctness of those reasons. This is how arguments for deliberative democracy typically proceed. These arguments are based on the idea that decision-makers should deliberate with one another by providing reasons that each can endorse. In this sense, fair procedures are also those that respect justification through deliberation.

Although there may be some conflict between these values, on the whole, these ideals all point in the same direction when we're thinking about fair procedures. These values suggest that it is necessary to give each actor equal respect and recognition in a decision-making process. They also suggest that decisions should be grounded in some notion of equality. Further, they all emphasise the agency of individuals, suggesting that decision-making processes should respect individuals as autonomous agents. Whilst there are different reasons for supporting fair procedures, for the most part, these all endorse the same approach to procedural justice.

This provides three intrinsic grounds for fair procedures. But these values aren't just important for the fairness of decision-making procedures in the UNFCCC. Designing decision-making procedures that respect autonomy, equality and justification brings about other, instrumental benefits, and it's worth briefly commenting on these.

For one thing, designing decision-making procedures that respect these values should also bring about decisions that accurately reflect the preferences and promote the interests of decision-makers. That is, decisions should also be more desirable on epistemic grounds. Some ways of making decisions, such as flipping a coin, fall down in this respect. But designing procedures that respect the values outlined above can bring about outcomes that have more epistemic value. For example, procedures that respect autonomy by giving people control over decisions are likely to lead to more accurate outcomes, since people are often best placed to know what their interests are. Alternatively, procedures in which people justify their decisions to one another are likely to be epistemically superior, because deliberation encourages people to learn about each other's interests (Christiano 1996, p. 84). It is therefore important that decision-making procedures accurately reflect the interests and preferences of a group as much as possible.

In addition to the epistemic accuracy of a procedure, solidarity is also important for designing procedures in the UNFCCC. Following the discussion from Chap. 3, it is not enough that the procedures of the UNFCCC are fair; decision-makers must *perceive* these procedures as fair. This is so that it can gain the necessary support for coordinating action on a large scale. But this means that other factors are also important for assuring decision-makers that their interests are sufficiently taken into account in a decision-making process. In this respect, solidarity is also important, in the sense that it encourages decision-makers to see the process as fair and it instils institutional norms of cooperation and collective action.

But this isn't all that's important; fair procedures are also those that pay attention to some substantive values. In Chap. 2, I argued that there is reasonable disagreement about some of the ends that the UNFCCC should pursue. But this does not imply that there is complete disagreement on this issue. In fact, it is likely that there is a great deal that decision-makers will agree about on what it should achieve, including issues such as the need to avoid dangerous climate change, and the need to avoid outcomes that violate people's basic human rights. The UNFCCC should certainly have procedures that avoid these sorts of ends, and the existence of reasonable disagreement doesn't refute this approach. Whilst decision-makers may disagree on some issues, it is likely that they will agree on some substantive ends. In this respect, there are also some minimum substantive ends that procedures should achieve.

Now that I've introduced the grounds of procedural justice, it is possible to think about what fair procedures look like in the UNFCCC. The remainder of this chapter provides some initial ideas about what this means for decision-making in general by defining a notion of political equality for procedural justice. It then shows how this notion of political equality should be interpreted in the UNFCCC.

5.3 Political Equality and Climate Change

Many people think that political equality is an important part of fair procedures. That is, there is something intrinsically valuable about the fact that decision-makers are equal in some respect. These arguments also arise in the context of climate change negotiations, where many have commented on the unfair disparity of resources that different delegations have for influencing decisions. In this section, I use the normative arguments developed in section two to think about what fair procedures look like in the UNFCCC.

In doing so, I argue that there are two principles of political equality for climate change institutions. The first principle is that, (1) decision-makers should have equal status in a decision-making process. In this case, equal status concerns both (1.1) equal status by the decision-making authority, and (1.2) equal status by other decision-makers.

But in addition to this, political equality implies a second principle; (2) decision-makers should have the equal opportunity to influence decisions. There are two subsidiary principles for achieving the equal opportunity to influence decisions: (2.1) decision-makers should have sufficient resources to participate on equal terms and (2.2) the rules of a decision-making process should be such that decision-makers face a level playing field. In what follows, I draw on the earlier arguments of this chapter to show why these principles are necessary for procedural justice. I then show how these principles can be achieved in the UNFCCC by making some practical policy recommendations that can serve as a guide for thinking about its procedural reform.

Turning first to the principle of equal status, here is something important about being treated and recognised as an equal in a decision-making process, regardless of what this means for one's ability to participate. To be sure, it's important that decision-makers *can* participate, in the sense that they have the ability to influence a decision. But this isn't the only thing that's relevant for procedural justice. It is also important that decision-makers are recognised and respected, in the sense that they are treated in a fair way, irrespective of what this means for their ability to participate. If a particular delegation is frequently provided with resources that are of poorer quality than those provided to other delegates, then there is something unfair in the way that they are treated. If, for example, a delegation is frequently placed at the back of the room, or is provided with a rest area that's of poorer quality than other delegations receive, then there's something unfair in the way that it's treated, even if it has no impact on their ability to participate in a decision.

An example that illustrates how the UNFCCC sometimes falls down on this issue concerns the language of negotiation documents. Sometimes, the documentation given to decision-makers is only provided in certain languages (UNfairplay 2011). Delegates whose native languages aren't included in these documents may still be capable of participating in decisions effectively, if these delegates speak the languages that are represented, or employ translators. But this still represents a matter of procedural justice, even if people can participate on equal terms. The point is that some actors are treated unequally by the decision to publish documents in specific languages.

The first principle of political equality appeals to this idea. In the UNFCCC, it is important, as a matter of fairness, that decision-makers receive equal standing and status. The 'status' that each decision-maker receives relates to how decision-makers are treated. Procedural justice requires that similar types of decision-makers should receive the same status. This means that *the same types* of decision-makers are treated alike and recognised as agents worthy of respect. It might be appropriate to treat different types of decision-makers differently. In particular, state delegations might be given more prominence and recognition than NSAs, given that they are typically seen as the most important actors in global politics by virtue of their democratic accountability or national sovereignty. What's more, it might be worth giving some NSAs different status to others (for example, it might be worth treating NSAs that represent indigenous groups differently to those that represent business interests). But similar decision-makers should at least have the same equal status in the decisions of the UNFCCC. This should be equal, because giving some more recognition necessarily means giving another less, and the important point here is that each decision-maker is treated alike.

This means that decision-makers are treated alike, to the extent that they are equal in other important respects.[8] This is similar to David Estlund's notion of anonymity, where anonymity means that, to the extent that each actor is equal, they should be treated as if they are interchangeable (Estlund 2008, p. 73). Equal status means that the rules do not single out any particular individual (provided that they are equal in every other sense) (Cohen 1997, p. 74). For Knight and Johnson, this means that procedures shouldn't discriminate between decision-makers (Knight and Johnson 1997, p. 288). There may be good reasons for treating actors differently if, for example, someone is unreasonable, or affected by a decision more than another actor.

One reason why this is important follows the discussion of status and solidarity that I gave earlier. It is not enough that decision-making processes are fair; decision-makers must *perceive* these processes as fair. This is only likely to happen if decision-makers are treated as equals in a decision-making process. This is a necessary condition for assuring each decision-maker that its views are taken into account in a decision-making process because decision-makers are unlikely to think that their views are treated equally if they are given less recognition or respect than others.

But equal status is also an important part of decision-making processes that respect autonomy, justification and equality. Giving each decision-maker the same status in a decision-making process instils norms of equality and inclusiveness among decision-makers. Each decision-maker realises that other decision-makers should be treated with the same respect and none should be privileged or marginalised in decisions. If the UNFCCC doesn't give these actors sufficient recognition in this respect, then it seems unlikely that it will in others. So equal status is also important for thinking that a decision-making process recognises our interests.

So there are several reasons for thinking that equal status is important for procedural justice, and these are based on the reasons that I explained earlier. Having

[8]Many attribute the idea that justice involves treating similar cases alike to Aristotle (Aristotle 1999, book 5).

explained why equal status is important, it's worth giving some greater definition to what this principle really implies. I now claim that equal status consists of two elements, each of which relates to the identity of the actor that should 'recognise' the decision-maker: (1) decision-making authority, and (2) other decision-makers.

5.3.1 The Decision-Making Authority

The decision-making authority of an institution can be thought of as the body that organises, coordinates and controls its decision-making process. In the UNFCCC this relates to the Secretariat, but it also relates to the COP Chair and its associated bodies, which can play a significant role in determining how decisions actually take place. The point is that there are certain actors that organise and coordinate decisions in this context and the way that these actors treat decision-makers is important for procedural justice.

Looking back at COP negotiations, it seems there's good reason for thinking that some decision-makers have been systematically marginalised. At COP15 in Copenhagen, for example, the delegation of Tuvalu submitted a detailed text proposing ambitious obligations for major developing countries. Despite strong support from civil society actors and several state delegations, the proposal was not seriously considered as a possible negotiation platform. The presiding chairpersons repeatedly ignored calls from several delegates to negotiate on the basis of Tuvalu's proposal. At one point after persistent demands to address the Tuvalu proposal, the Chair responded "Tonight when I go to bed, I will take the proposal with me for bedtime reading."[9] Radoslav Dimitrov claims that other delegations wouldn't have been treated in this way and that this represented a lack of respect to the delegation of Tuvalu (Dimitrov 2010b, p. 807). This shows the sort of marginalisation that some decision-makers face from the decision-making authority itself. It is not just the fact that Tuvalu's proposal wasn't taken into account; what is also important is the way that the decision-making body treated the proposal. This gives an illustration of how the decision-making authority may treat decision-makers unfairly in the UNFCCC.

Following the discussion earlier, a decision-making authority should give actors equal status in a decision. There are several measures that the UNFCCC can take to ensure that the decision-making authority treats each decision-making with equal status. For example, as I already suggested, information should be provided in a way that recognises the equal worth of each decision-maker. This might mean providing documents in every native language of the states present in the negotiation forum. Other measures might concern the location of institutional conferences. The COP to the UNFCCC meets biannually, and the major party conference takes place in a different location each year, with the host nation acting as the Chair for the event.

[9] See: Dimitrov 2010b, p. 807.

The UNFCCC could ensure that decision-makers are treated as equals by allowing each state to host the party conference at some stage, rather than limiting meetings to certain states. Other issues might include ensuring that decision-makers should have equal opportunities to address the decision-making forum and raise issues for discussion.

5.3.2 Other Decision-Makers

Having looked at how the decision-making authority should treat decision-makers, it's now necessary to consider the second element of equal status: equal recognition from other decision-makers. The problem is that equal status doesn't just apply to how a decision-making authority treats decision-makers; how decision-makers interact with each other is also a concern.

To see this, it's worth considering some of the ways that decision-makers have interacted with each other in the UNFCCC in the past. On at least some occasions there are clear instances where decision-makers haven't treated each other sufficient respect. For example, during the closing stages of COP15 in Copenhagen, a 'Friends of the Chair' group was created in response to the procedural problems of drafting an agreement amongst so many actors.[10] This group consisted of the heads of state of 25 countries whose purpose was to draft an agreement. This largely excluded many other delegates and negotiators who were left out of the drafting process. Later, the heads of state of the United States, China, India and Brazil met in private in order to finalise the details of this, producing what became known as the Copenhagen Accord. Before the entire COP delegation had a chance to read this text, the members of this exclusive group held a press conference claiming that a global deal had been achieved.[11] This sparked outrage from the broader COP delegation, many of which weren't just kept of out these discussions, but were wholly unaware that they had taken place (Dubash 2009). As a result, Venezuela, Cuba, Nicaragua and Bolivia later renounced the agreement on the grounds that this exclusive group was trying to impose an agreement on the rest of the COP.

So equal status isn't just about respect from a decision-making authority; it also concerns fair treatment by other decision-makers. Following the example above, fairness means that actors should treat other reasonable actors (and their claims) with respect. Recalling what was said in Chap. 2, fair decisions are those in which actors participate in a reasonable way, where reasonableness puts certain constraints on the types of claims that an actor can make. These requirements stipulate, for example, that claims should be based on sound reasoning and that they have some normative basis. This essentially means that reasonable actors are those who put forward views that other reasonable actors are willing to accept

[10]See: Dimitrov 2010b, p. 809–10.

[11]See: Dimitrov 2010b, p. 810.

as reasonable claims. But this doesn't just mean that actors make reasonable claims; reasonableness also means actors treat other reasonable claims with due respect.

Equal status also entails that participants should not engage in manipulation or coercion. In terms of bargaining strategies, this means adopting soft, or cooperative strategies, as opposed to hard, or aggressive ones.[12] For Christiano, this is a necessary feature of treating people as equals. If an outcome is determined through manipulation then it doesn't treat others as equals, because doing so means that some prioritise their own ends and interests above those of others (Christiano 2008, p. 99). Equal status is therefore necessary to achieve the equal advancement of interests. Given that manipulation involves controlling another actor through deceit, equal recognition is also important for autonomy.

This in turn means that decision-makers should engage with each other under a notion of reciprocity. That is, actors should not merely set out to maximise their own self-interest; they should show some restraint in their actions. But this doesn't mean that they act entirely selflessly either. Rather, they operate at some midpoint between these positions. This requires that there be an absence of threats, coercion and manipulation, which are issues that I discuss in greater detail in Chap. 6. For the moment, it is sufficient to define equal status as the equal recognition of decision-makers, where this applies to (1) the institution, and (2) other decision-makers.

This represents an account of what's required for equal status. But this isn't all that matters for political equality. Each decision-maker might have equal recognition in a decision-making process whilst important differences between them are ignored. For this reason, it is also necessary to consider the capability of each actor to participate in a decision. Procedural justice requires that decision-makers have the equal opportunity to influence the decision-making processes of the UNFCCC.

5.4 Political Equality and Equal Opportunity (1): Resources and Capabilities

Political equality is not just about equal status; decision-makers should also have an equal opportunity to influence decisions. This follows from the arguments in section two. Decision-makers should be able to advance their interests equally in a decision-making process in a way that allows them to act as independent agents. This means that decision-makers should have the same opportunity to influence a decision, where influence concerns the amount of control that a decision-maker has over the outcome of the decision.

There is an on going debate, however, regarding the means that should be used to achieve equal opportunity of influence in decisions. This debate concerns

[12]For discussion, see: Weiler 2013.

whether it's best to think of political equality in terms of equal resources or equal capabilities. In what follows, I discuss each of these views, before arguing that it is best to achieve equal opportunity for influence in the decision-making processes of the UNFCCC by equalising actors' *capability* to participate.

Take the resources approach first. John Rawls argues that it is possible to achieve equality of opportunity for political influence by focusing on the resources that each actor has for participating in a decision (Rawls 1993, p. 183). According to Rawls, it is possible to provide an equal opportunity for participation in political decisions by giving people sufficient 'primary goods', where primary goods relate to the rights and liberties that people use to influence political decisions. In COP negotiations, these sorts of resources, or 'primary goods', might concern the voting rights that each state delegation has, the amount of time that it gets to speak in a debate, or its right to participate in discussions. Part of the desirability of Rawls' approach is that it is simple and comparatively easy to implement; each decision-maker is given an equal amount of some defined resource. Although Rawls recognises that citizens differ in their ability to convert these resources into political influence, he assumes that each actor has the basic capability to participate as cooperative members of society.

But the problem with this approach is, as Rawls notes, that it neglects important differences in people's ability to use these resources to influence decisions. In some situations, this might not be so problematic. If decision-makers are largely alike, and if none can gain a sufficient advantage over another in terms of influence, then equalising the resources that each decision-maker has for making decisions may be an effective way of realising equal opportunity of influence. But it is possible to imagine that, where decision-makers differ very widely in this respect, this is a significant problem for achieving equal opportunity for political influence. If, for example, some decision-makers are more charismatic or eloquent than others, then giving each decision-maker the same amount of time to speak in a debate may leave others with less influence over a decision.

For this reason, Knight and Johnson draw on James Bohman's work to propose a capabilities approach for deliberative democracy, which focuses on the freedoms that people have to use resources for political functionings (Bohman 1997; Knight and Johnson 1997). This is provides a useful starting point for thinking about political equality in decisions more generally. Capabilities are the various capacities or abilities that actors have to perform a certain function, where a function is an activity that an agent can undertake.[13] Rather than focussing on the resources, or goods that a person has, the capabilities approach focuses on what people can achieve with these goods. Giving a decision-maker the right to participate in a decision is only part of what is required for giving each decision-maker the equal opportunity to influence an outcome. If decision-makers are to be able to advance their interests equally and independently then it is necessary to think about the different abilities that people have to make use of this right. Because capabilities concern the *ability* of an

[13] Amartya Sen advocates the use of a capabilities approach in relation to welfare (Sen 1993).

actor to perform a particular function, equalising capabilities is a more appropriate way of achieving the equal opportunity for political influence when there are large discrepancies in people's ability to convert decision-making resources into political influence. It recognises that there are sometimes fundamental differences between decision-makers that prevent them from influencing decisions to the same extent as others.

On the one hand, equalising resources seems the most appropriate means for ensuring equal opportunity of influence in a decision where there are not significant differences in the ability to convert these resources into influence, or when capabilities are hard to define. On the other hand, equalising capabilities is more appropriate for this end when there are significant differences between actors.

Turning to the issue of decision-making in the UNFCCC, the literature on this topic suggests that decision-makers do differ significantly in their ability to participate in its decision-making processes. Some state delegations, in particular those that represent wealthy or populous states, often have access to greater technical and legal expertise than others.[14] Whilst some can afford large numbers of highly skilled delegates and negotiators, others only have the resources to send a limited number to conferences (Schroeder et al. 2012). This means that some state delegations are at a distinct disadvantage in their ability to influence decisions in the UNFCCC. Even when these delegations receive the same primary goods, such as the right to participate or the right to speak in debates, the large asymmetries between these actors mean that some can influence a decision to a far greater extent than others. Given that it is best to think of political equality in terms of equal capabilities when there are large differences in the ability of decision-makers to participate in decisions, it is most appropriate to achieve equal opportunity of political influence in the UNFCCC by ensuring that each decision-maker has the equal *capability* to participate in a decision. Unlike an approach that focuses on equalising resources, this captures the concern that different decision-makers can convert decision-making resources to different extents.

5.5 Political Equality and Equal Opportunity (2): Capabilities and Functionings

Given that it is best to think of political equality in terms of equal capabilities in the UNFCCC, it's now necessary to think about what sorts of capabilities are most relevant here. Several authors have discussed the relevant capabilities that are necessary for participating in decisions generally, where the relevant function here is 'to participate in a decision-making process'.[15] To give this notion of

[14]See, for example: IPCC Special Report on Participation of Developing Countries (IPCC 1990).

[15]See, for example: Bohman 1997; Knight and Johnson 1997.

participation some depth, Knight and Johnson identify several necessary features that are relevant for participation, including the 'capacity to formulate authentic preferences', 'the effective use of cultural resources,' and, 'basic cognitive abilities and skills' (Knight and Johnson 1997, p. 298). Although this provides a useful staring point for thinking about functions in decision-making contexts, this list is primarily concerned with deliberative decisions in a domestic democracy. Given that our concern is the decision-making processes of the UNFCCC, it seems reasonable to suppose that the decision-makers in these institutions (i.e. delegations) already have the basic cognitive ability to participate in decisions and it doesn't seem so important to address these issues here. Rather, in what follows, I propose a more relevant account of the functionings that are important for participating in the decision-making processes of the UNFCCC. These are: forming opinions, making judgements, and advocating interests and positions.

Decision-makers should have the capability to formulate views and make opinions about decisions. This involves understanding the issues involved, interpreting information, and making judgements about what positions to adopt. This doesn't just require that decision-makers *can* form views and opinions. After all, decision-makers may form opinions that are based on incorrect information, or on poor judgement. Rather, this means that actors can form accurate opinions, which are based on correct information and sound reasoning.

Decision-makers should also be able to influence the outcome of a decision. In some situations, this means that decision-makers should have a right to vote. But, as I argued in Chap. 4, in some situations decision-makers should be able to influence a decision-making process by having a voice, but not a vote in the decision. In this latter situation, if an actor is to effectively contribute to a debate and to influence an outcome then the ability to form judgements and opinions is insufficient for participation. Decision-makers should also be able to put forward their views and to have these heard in debates and discussions.

This means two things. First, decision-makers should be able put forward their views and to express these views effectively in a debate. Second, decision-makers should be able to *influence* the decision-making process, in the sense that they can persuade other decision-makers and have some control over the outcome of a decision. This doesn't necessarily equate to having a vote, or a say in a decision-making process. Rather, this concerns the ability to change peoples' minds and to advocate an argument effectively.

In conclusion, procedural justice requires that decision-makers have the equal opportunity to influence a decision-making process, and in the UNFCCC this means that decision-makers should have the equal capability to participate in these decisions, where capabilities concern the ability to achieve to some functionings. In what follows, I argue that two principles are necessary for achieving this end. One relates to resources, whilst the other concerns the rules that decision-makers face.

5.6 Political Equality and Equal Opportunity (3): Sufficient Resources

If the UNFCCC should ensure that each decision-maker has the equal capability to participate, then this requires that decision-makers should have enough resources so that they have the equal opportunity to influence a decision, where influence relates to the ability to make informed judgements about issues, to put forward views, and to influence others.

To substantiate this claim, it is necessary to address three questions. What sorts of resources are important for ensuring that decision-makers have the capability to participate in decisions? Why is sufficiency the relevant standard here? And what is an adequate baseline for a sufficient distribution of these resources in the UNFCCC?

Let's start by considering the question of resources first. Certain resources are important because they enable decision-makers to influence decisions. Given our account of procedural justice in section two, decision-makers need these resources in order to participate on equal terms. Some authors describe these resources as either internal or external.[16] Internal resources are those that are inherent to the decision-making process itself, including, for example, information about the decision, one's aptitude at persuasion, and negotiation or bargaining skill. External resources, on the other hand, are resources that are not directly linked to the decision-making process, but nevertheless have some bearing on the ability of an actor to participate. External resources include things such as material wealth or physical power. Whilst some authors suggest that it is necessary to focus on both of these issues, the UNFCCC should concentrate on internal decision-making resources.

For one thing, it's prohibitively costly to change the distribution of external resources that decision-makers have in the UNFCCC. At the multilateral level, external resources concern things like global political clout and economic GDP. Not only would it be highly impractical to equalise some of these resources, in other cases it would place demands on actors that wouldn't be reasonably feasible in practice. Changing the external resources that state actors have is an unduly extreme measure for achieving political equality in a decision-making process that seems unlikely to ever arise in a practical sense. Equalising external resources focuses on the wrong issue. What's more important is ensuring that disparities in these resources do not play a strong role in determining the influence that each decision-maker has in a decision.

In a multilateral context, this means that it is necessary to focus on ensuring that decision-makers have sufficient *internal* resources for decision-making instead. These are resources that determine how much a decision-maker can influence a decision, including the amount of information that decision-makers have, the

[16]For example: Weiler 2013.

technical expertise for interpreting information and forming opinions and positions, and the resources for advocating positions in a decision-making process, such as the number of delegates that each actor has.

Turning to the issue of sufficiency, ensuring that each decision-maker has an equal capability to participate in the UNFCCC requires a sufficient, rather than an equal, distribution of internal resources. This is because procedural justice requires that decision-makers can participate on equal terms. But the problem is that some decision-makers face disadvantages on account of the fact that they do not have enough resources to participate effectively. This means that the UNFCCC should ensure that each decision-maker has sufficient resources to participate on equal terms.

A second reason for thinking that sufficiency is the relevant standard for this context comes from the earlier discussion of capabilities. Some decision-makers can make use of resources to different extents. For example, if a particular decision-maker has access to high quality legal and technical expertise then it may be able to use other resources, such as information, to better use than another decision-maker that isn't in such an advantageous position. This means that it is necessary to ensure that each decision-maker has *sufficient* resources for influencing an outcome. That is, given that some use these resources more effectively than others, and given that each should participate to the same extent as others, it is important to ensure that each has enough to participate. But this doesn't require ensuring that each decision-maker has *equal* resources. What matters is the ability of each decision-maker to participate, rather than equality per se.

One might object that a problem with this approach is that some resources are positional. A positional good is valuable to the extent that an actor has more of it than someone else. In the UNFCCC, this might be the case with the number of delegates that each state delegation can send to a meeting. It is often noted that developing country delegations are much smaller than those from developed nations in multilateral negotiations and that this presents serious challenges to the ability of developing countries to negotiate on equal terms.[17] Once a delegation has enough delegates to participate in a meeting, the number of delegates is not valuable in itself; rather, it is the fact that some states have more than others that gives them an advantage. This might lead us to think that the UNFCCC should ensure that each has *equal* internal resources, rather than a sufficient distribution as I've proposed here.

This argument does not, however, show that what really matters here is equality. Rather the focus should be on giving each decision-maker sufficient resources to participate. It is just the case that where there are positional goods, giving each decision-maker sufficient resources to participate on equal terms also means giving each decision-maker equal resources.

Of course, this doesn't mean that the UNFCCC should allow decision-makers to accumulate unlimited resources for decision-making. This is particularly relevant

[17] For various accounts of this, see: Bruce et al. 1995; South Centre 2004; UNfairplay 2011.

for things like delegation size, where the number of delegates that each state brings to negotiations has an impact on the overall quality of the decision-making process. The UNFCCC should put limits on these sorts of resources for the sake of practicality. The appropriate criterion should be whether bringing more resources to the table is detrimental for the decision-making process.

Turning to the question of what a suitable baseline of resources is, decision-makers should have a sufficient amount of resources to participate as equals in a decision-making process, *given the design of the decision-making process and given the differences between them.* This means avoiding situations where decision-makers are unable to participate, or forced to compromise on an issue because they have insufficient resources. For example, some suggest that resource disparities have forced some participants to depend on coalitions in order to be heard in COP negotiations, altering their negotiation positions in order to find common ground with others (Gupta 2000; Chasek and Rajamani 2003). Others argue that insufficient resources have also led decision-makers to adopt reactive, defensive, and negative negotiating positions (Richards 2001). In other cases, some states do not have sufficient resources to attend different multilateral negotiations, and are forced to choose where they will prioritise their interests.[18] These are clearly cases where states have insufficient resources to participate and it is necessary to avoid situations where these issues arise.

Given that each should have sufficient resources to participate on equal terms, it's also necessary to consider which agents are responsible for providing these resources. Since fair procedures are those in which decision-makers maintain some degree of autonomy and independence, each decision-maker should be responsible for providing their own resources for participation. But the point is that some actors have insufficient resources to participate on equal terms. There are several measures that the UNFCCC itself can take to ensure that each actor has access to a sufficient amount of resources for equal capability for participation.

First, NSAs play an important role. NSAs improve the negotiating position of state delegates by providing additional expertise and assistance to smaller state delegations.[19] NSAs often play an epistemic role in this respect by improving the information available to decision-makers. These ideas are supported in much of the literature on multilateral negotiations. For instance, UNfairplay is a policy brief created to augment the negotiating capacity of countries with small delegations by encouraging their engagement with civil society actors (UNfairplay 2011). The potential role that NSAs can play in assisting developing country delegates in multilateral negotiations is also recognised by Chasek and Rajamani, as well as the South Centre (Chasek and Rajamani 2003; South Centre 2004). These studies show how NSAs can help to provide decision-makers with the necessary resources to participate effectively in multilateral negotiations.

---

[18]For example, Schroeder et al. argue that differences in delegation size in climate institutions partly reflect the priority that states with limited resources give an issue (Schroeder et al. 2012).

[19]For support of this claim, see: Steffek et al. 2008; Scholte 2011.

A second option is to pool decision-making resources among state delegations. Some actors share many interests in multilateral negotiations, and forming coalitions or alliances is one way of improving the resources that these actors have for advocating these interests.[20] Radoslav Dimitrov notes that there are several coalition groups in the UNFCCC that help states to advocate their interests in its decisions (Dimitrov 2010a). These allow state delegates to combine resources and improve their joint ability to participate in decisions. The UNFCCC can promote political equality by encouraging states with similar interests and negotiating positions to engage with one another. At the same time, it is necessary to keep in mind this might force people to compromise their positions for the sake political influence. Whilst coalitions can bring about a number of benefits, this shouldn't come at the cost of compromised values.

A third measure is to promote domestic institutions as enabling agents that can support actors in COP decisions.[21] Decision-makers should act as effective representatives of their constituents, as well as of other stakeholders. These decision-makers should also be able to participate effectively in decisions, which requires a sufficient standard of negotiation resources such as information and decision-making ability. But this requires a suitable amount of institutional development on the part of the actors that are participating in these decisions. That is, it's necessary to have a sufficient amount of education and awareness at the domestic level, which can translate into effective representation at the multilateral level. For this reason, the negotiation capability of state actors can be improved through improvements in domestic institutions. This is something that is supported in the literature on negotiation capacity. For instance, the UNfairplay report notes that domestic capacity is a crucial feature of negotiation process (UNfairplay 2011, p. 12). In terms of trade negotiations, the South Centre notes that this may require countries to make significant resource investments in the development and improvement of the human, technical, financial, and physical infrastructure (South Centre 2004, p. 2). Improving the domestic institutions of states is therefore a third way of helping states to participate under terms of political equality.

5.7 Political Equality and Equal Opportunity (4): A Level Playing Field

Political equality isn't just about providing actors with sufficient resources to participate on equal terms. The problem is that the ability to participate in a decision is not just a matter of resources; it is also a matter of the procedural rules that decision-makers face. Access to sufficient resources isn't enough for equal opportunity of influence; it's also necessary to ensure that decision-makers face a level playing

[20] For discussion: Gupta 2000; Chasek and Rajamani 2003.

[21] Note that Chasek and Rajamani 2003 also make this argument.

field. For this reason, it's also necessary to think about the rules of the decision-making process and how these affect an actor's ability to participate effectively.

This requires a second principle for achieving equal opportunity of influence, which is that the rules of the decision-making process should be such that each decision-maker faces a level playing field. That is, procedural justice requires that the rules of decision-making process are designed so that, given that each has a sufficient amount of resources for participation, each actor is able to participate to the same extent as everyone else. Following the discussion in section three, each decision-maker should have equal status in a decision-making process. But in addition to this, the rules of the decision-making process should allow decision-makers to participate on equal terms. Even if there are asymmetries between parties in a decision-making process (for instance, in external resources, or negotiation skill) these should not be allowed to influence the ability of decision-makers to participate in a decision-making process, given our understanding of participation as the ability to form opinions, make judgements, and influence an outcome. The rules of the game should allow each decision-maker to participate to the same extent as others, given its background resources. These rules relate to the design of the decision-making process. This concerns rules such as the amount of time that each view gets in a discussion, the amount of information necessary to make an informed judgement on a decision, the language used in a decision-making process, or the number of meetings that each actor has to attend.

In order to realise what this really means, it's helpful to draw from some of the commentary from recent COP negotiations. Several authors note that marathon negotiation sessions that overshoot the scheduled completion of the conference are now a common characteristic of COP negotiations.[22] This is extremely demanding for states that are only able to provide a handful of delegates, or that are limited by financial constraints. For example, when the final negotiation session of COP17 overran by over 36 h, several delegates had to leave the conference early to avoid missing their flights home.[23] This was after a single negotiation session had spanned three consecutive nights. Another issue concerns the large number of issues and agendas that a negotiation process attempts to address. Small delegations can struggle with the heavy workloads that are squeezed into one negotiation session, or the number of negotiation bodies that run simultaneously during a negotiation. For example, in the UNFCCC there are six negotiating bodies that meet at the at the annual COP negotiations.[24] These examples give us some understanding of the problems that decision-makers face, as a result of the rules of the institution itself.

The UNFCCC should take measures to resolve these issues by ensuring that each actor faces a level playing field in the negotiation process. It can achieve this end by implementing specific procedural rules for ensuring that its decision-making processes don't marginalise any delegations. Many environmental institutions

[22]For example: Werksman 1999, p. 13; Harvey and Vidal 2011.

[23]See: Harvey and Vidal 2011.

[24]Vihma and Kulovesi 2012, p. 7. See also: Doran and Gloel 2007; UNfairplay 2011, p. 10.

now have rules that to facilitate the participation of certain actors (in particular, developing countries with small delegations). In the UNFCCC, there is already a requirement that all documentation is published simultaneously in all six UN languages so as to make it possible for a larger majority of delegates to participate fully in the negotiations.[25] Similar measures that promote equality include limiting the number of meetings that take place simultaneously, so that small delegations are able to attend each of them.

The Chair of the COP can play a major role in this respect. The Chair changes according to where each COP of the UNFCCC is hosted. But it plays a large role in interpreting how decisions actually take place, determining, for example, the amount of time that each actor gets to speak, who gets to speak when, and how long a negotiation session can run for. These are all important issues for making sure that each actor faces a level playing field in negotiations. The Chair of each COP should have a specific mandate to ensure that this happens.

5.8 Conclusion

Procedural justice is about more than providing the right to participate in a decision-making process. It also requires that actors can participate effectively, and on fair terms. In this chapter, I've argued that procedural justice is grounded in values of autonomy, equality and justification. I've also argued that this means that there is a standard of political equality for fair decision-making in the UNFCCC. This standard should be understood as an equal status in decision-making processes as well as the equal opportunity to influence the outcome of a decision-making process. I've also made some suggestions for how this standard can be achieved in practice. What this chapter hasn't addressed is how actors should bargain in a decision-making process, or how much influence (in terms of a say) each actor should have according to procedural justice. These are important issues in their own right, which I consider in Chaps. 6 and 7.

References

Aristotle. 1999. *Nicomachean ethics*. Indianapolis: Hacket

Beitz, C.R. 1989. *Political equality: An essay in democratic theory*. Princeton: Princeton University Press.

Bohman, J. 1997. Deliberative democracy and effective social freedom: Capabilities, resources, and opportunities. In *Deliberative democracy: Essays on reason and politics*, ed. J. Bohman and W. Rehg. Cambridge, MA/London: MIT Press.

[25]This is a more specific interpretation of draft responsibilities echoed in the UNFCCC founding text: [the UNFCCC Secretariat shall] 'receive, translate, reproduce and distribute the documents of the session' (UNFCCC Draft Rule 29b).

Bruce, J.P., L. Hoesung, et al. 1995. *Climate change 1995: Economic and social dimension of climate change. Contribution of Working Group III to the Second Assessment Report of the Intergovernmental Panel on Climate Change*. Cambridge: Cambridge University Press.
Chasek, P., and L. Rajamani. 2003. Steps toward enhanced parity: Negotiating capacities and strategies of developing countries. In *Providing public goods: Managing globalization*, ed. I. Kaul, P. Conceição, K. Goulven, and F. Mendoza. Oxford: Oxford University Press.
Christiano, T. 1990. Freedom, consensus, and equality in collective decision making. *Ethics* 101(1): 151–181.
Christiano, T. 1996. *The rule of the many: Fundamental issues in democratic theory*. Boulder/Oxford: Westview Press.
Christiano, T. 2004. The authority of democracy. *Journal of Political Philosophy* 12(3): 266–90.
Christiano, T. 2008. *The constitution of equality: Democratic authority and its limits*. Oxford: Oxford University Press.
Cohen, J. 1989. Deliberation and democratic legitimacy. In *The good polity: Normative analysis of the state*, ed. A. Hamlin and P. Petit. New York: Blackwell.
Cohen, J. 1997. Procedure and substance in deliberative democracy. In *Deliberative democracy: Essays on reason and politics*, ed. J. Bohman and W. Rehg. Cambridge, MA/London: MIT Press.
Dahl, R.A. 1956. *A preface to democratic theory*. Chicago/London: University of Chicago Press.
Dimitrov, R.S. 2010a. Inside Copenhagen: The state of climate governance. *Global Environmental Politics* 10(2): 18–24.
Dimitrov, R.S. 2010b. Inside UN climate change negotiations: The Copenhagen conference. *Review of Policy Research* 27(6): 795–821.
Doran, P., and J. Gloel. 2007. *Capacity of developing countries to participate in international decision-making*. International Institute for Sustainable Development.
Dryzek, J. 2000. *Deliberative democracy and beyond*. Oxford: Oxford University Press.
Dryzek, J.S. 2010. *Foundations and frontiers of deliberative governance*. Oxford: Oxford University Press.
Dubash, N.K. 2009. Copenhagen: Climate of mistrust. *Economic & Political Weekly* XLIV(52): 8–11.
Estlund, D.M. 2008. *Democratic authority: A philosophical framework*. Princeton/Oxford: Princeton University Press.
Gould, C. 1988. *Rethinking democracy: Freedom and social cooperation in politics, economics and society*. New York: Cambridge University Press.
Gupta, J. 2000. *"On behalf of my delegation ... " A survival guide for developing country climate negotiators*. Published jointly by the Center for Sustainable Development of the Americas and the International Institute for Sustainable Development.
Gutmann, Amy, and D.F. Thompson. 2000. Why deliberative democracy is different. *Social Philosophy and Policy* 17(01): 161.
Harvey, F., and J. Vidal. 2011. *Global climate change treaty in sight after Durban breakthrough climate conference ends in agreement after two weeks of talks*. London: The Guardian.
IPCC. 1990. IPCC special report on participation of developing countries. *Intergovernmental Panel on Climate Change*.
Kelsen, H. 1955. Foundations of democracy. *Ethics* 66(1): 1–101.
Knight, J., and J. Johnson. 1997. What sort of political equality does democratic deliberation require? In *Deliberative democracy*, ed. J. Bohman and W. Rehg. Cambridge, MA: MIT Press.
Mansbridge, J., et al. 2010. The place of self-interest and the role of power in deliberative democracy. *Journal of Political Philosophy* 18(1): 64–100.
Rawls, J. 1993. *Political liberalism*. New York: Columbia University Press.
Richards, M. 2001. *A review of the effectiveness of developing country participation in the climate change convention negotiations*. London: Overseas Development Institute.
Rousseau, J.-J. 1762. *The social contract*. Harmondsworth: Penguin.
Saunders, B. 2010. Democracy, political equality, and majority rule. *Ethics* 121: 148–177.

Scholte, J.A. 2011. *Building global democracy? Civil society and accountable global governance*. Cambridge: Cambridge University Press.
Schroeder, H., M.K. Boykoff, et al. 2012. Equity and state representations in climate negotiations. *Nature Climate Change* 2: 834–836.
Sen, A. 1993. Capability and well-being. In *The quality of life*, ed. M.C. Nussbaum and A. Sen. Oxford: Oxford University Press.
South Centre. 2004. *Strengthening developing countries' capacity for trade negoitations: Matching technical assistance to negotiating capacity constraints*. Written at the request of the Office of the G77 and China in New York for the G-77 and China High-Level Forum on Trade and Investment, Doha, Qatar, December 5–6, 2004.
Steffek, J., C. Kissling, et al. 2008. *Civil society participation in European and global governance: A cure for the democratic deficit?* Basingstoke: Palgrave Macmillan.
UNfairplay. 2011. *Leveling the playing field. A report to the UNFCCC secretariat on negotiating capacity*. http://unfccc.int/files/conference_programme/application/pdf/unfairplayreportapril202011-1.pdf.
Verba, S. 2003. Would the dream of political equality turn out to be a nightmare? *The American Political Science Association* 1(4): 663–679.
Vihma, A., and K. Kulovesi. 2012. *Strengthening global climate change negotiations; Improving the efficiency of the UNFCCC process*. Nordiske Arbejdspapirer Nordic Working Papers. Nordic Council of Ministers.
Weiler, F. 2013. Determinants of bargaining success in the climate change negotiations. *Climate Policy* 12(5): 552–574.
Werksman, J. 1999. *Procedural and institutional aspects of the emerging climate change regime: Do improvised procedures lead to impoverished rules?* Concluding Workshop for the Project to Enhance Policy-Making Capacity Under the Framework Convention on Climate Change and The Kyoto Protocol. Foundation for International Environmental Law and Development, London.

Chapter 6
Fair Bargaining; Voluntariness and Reciprocity

6.1 Introduction

Given the number of different parties and interests present, deliberation rarely brings about consensus from every member state in COP negotiations. In fact, as I showed in Chap. 2, there are certain situations where there is significant reasonable disagreement over the fair terms of cooperation within this forum. Given that climate change is a problem caused by actors around the world, adequately addressing this problem will depend on finding an adequately acceptable agreement amongst states on a universal scale. Given that the nature of global politics means that states must participate voluntarily, this raises the question of how the UNFCCC can proceed in the face of persistent reasonable disagreement.

There are at least two kinds of response. One approach is to use a voting method to arrive at a mutually acceptable agreement, which I discuss in Chap. 7. A second approach is to bargain, which involves making concessions or compromising on an issue for the sake of reaching agreement.[1] Given the difficulty of reaching universal agreement among large numbers of actors, bargaining is seen as critical part of effective decision-making in many multilateral institutions.[2]

Yet bargaining introduces problems for fairness. For example, when two parties value a negotiated agreement differently, bargaining might result in one actor being exploited. Some suggest that this happened in the preliminary negotiations of the UNFCCC, when many oil producing states put a relatively low valuation on reaching a negotiated agreement on climate mitigation (Depledge 2008). These states were subsequently able to hold out for large concessions from those who valued a negotiated outcome much more. In other cases, people often feel that there are limits to the types of bargains that it's possible to make in the UNFCCC.

[1] I take this definition from: Raiffa 1982, p. 142; for a similar definition, see Nash 1950.

[2] See: Sebenius 1984; Susskind and Ozawa 1992; Zartman and Touval 2010.

L. Tomlinson, *Procedural Justice in the United Nations Framework Convention on Climate Change*, DOI 10.1007/978-3-319-17184-5_6

Some claim that Russia's endorsement of the Kyoto Protocol was dependent on its acceptance into the World Trade Organisation (WTO). In this way, a mutually acceptable agreement was reached by 'linking' one issue to another. But some linkages, such as those involving side-payments, or those that relate to development aid, are often criticised as unfair (Albin 2001). It's therefore also necessary to think about what sort of linkages should be permissible in the UNFCCC.

Whilst bargaining is an important part of decision-making, fairness requires some limits on the extent to which parties can pursue their own self-interest in bargaining processes. Recalling the arguments from Chap. 5, fair decisions are those that promote autonomy and independence, allow actors to advance their interests equally, and encourage justification through deliberative procedures. In this chapter, I argue that this has certain implications for fair bargaining. I claim that fair bargains are those that allow people to advance their own interests in a decision, according to some constraints about what's reasonable. In specific, I argue that fair bargains are those that are (i) voluntary, and (ii) reciprocal. Bargains are voluntary if bargainers are informed and rational, and if bargains are free from coercion and manipulation. Bargains are reciprocal to the extent that they involve roughly equal concessions between bargainers.

Following these claims, I argue that there are several policy measures that can improve the fairness of bargaining in the UNFCCC. Foremost, I argue that there should be limits of the linking of certain issues. Given that many authors argue that issue linkage improves the likelihood of arriving at a mutually acceptable agreement, these conclusions represent an alternative approach to traditional policy recommendations for bargaining.

6.2 Voluntariness

Fair bargains are those that actors agree to voluntarily. That is, fair bargains are those in which actors enter freely and through their own volition. If a bargain is involuntary, then it is clear enough that the bargainers lose some elements of autonomy and independence. Consent is therefore a necessary feature of autonomy (Wertheimer 1989; Zwolinski 2007). But consent isn't enough for voluntariness. People often think that an arrangement is involuntarily if an actor is forced into an agreement, or if an actor is in no position to reasonably reject it, regardless of consent. To this end, it's worth spelling out some of the necessary conditions for voluntariness.

People usually think that an agreement is only voluntary if all parties are rational and fully informed about the bargain. This is a necessary, but not sufficient condition for fair voluntariness. For example, Serena Olsaretti argues that voluntary agreements are those in which the parties are fully informed about the terms of an agreement (Olsaretti 2004). This means that parties are aware of what the bargain involves, what is being exchanged, and what the value of the bargain is. People cannot make effective decisions unless they are informed about the decision and they cannot act autonomously unless they know what is at stake. Decision-makers should also be rational, in the sense that they can interpret information and make correct judgements about their interests. These conditions are also true for equality and justification in a decision-making process. Decision-makers can only advance their interests in a decision and justify their decisions to one another if they are aware of the issues at stake and are informed about the decision.

But whilst decision-makers need information about what is being exchanged, this doesn't extend to information about how much each actor values something. How much an actor values a good is often referred to as an actor's 'reservation price', which reflects the highest cost that an actor is willing to pay for a good. This is a critical feature of bargaining. If I know how much an agent values a good then, then I can demand the agents pays its full reservation price for that good. Procedural justice doesn't require that decision-makers know what each other's reservation prices are. House buyers are not obligated to tell estate agents how much they are willing to pay for a house, nor are bidders expected to reveal how much they are willing to pay for a lot in an auction. Hidden reservation prices are compatible with fair procedures, because information about these prices isn't necessary for voluntary bargains. Voluntariness only requires that each bargainer knows what is being exchanged and is able to make an accurate judgement of his or her own valuation of the good.

But voluntariness also requires a certain degree of rationality. This means that actors bargain under some minimal conditions of rational thought and that they are capable of acting according to their own interests. That is, people are rational if they are able to formulate opinions and advance their interests and views in a bargaining process through negotiation and discussion. This gives us two preliminary conditions for voluntariness: information and rationality. We can, however, go further. Voluntariness also involves being able to act independently of others. This gives us two further conditions, namely: an absence of manipulation, and an absence of coercion, which I discuss in turn below.

6.2.1 *Manipulation*

A further condition of voluntariness is that an agreement is free from manipulation, *if actors are reasonable*. There may be some cases where manipulation is justified if actors are unreasonable, but manipulation is unjustified if actors are reasonable. Manipulation, as with coercion and persuasion, is an attempt to influence someone's behaviour (Rudinow 1978). Where coercion attempts to change behaviour through threats, persuasion does so through the force of better argument. Manipulation, on the other hand, involves changing someone's behaviour either through deception or by taking advantage of a weakness. The problem with this, as I show below, is that it alters our volition.

There are many different accounts of what manipulation involves and why it is undesirable.[3] In general, these accounts share the premise that manipulation involves influencing another actor's behaviour, although they make different conclusions about how this is done. For Derk Pereboom, manipulation is about controlling an agent's actions in some way (Pereboom 2001, p. 112). But manipulation isn't just about controlling someone, in the sense of forcing someone to do something. Rather, for Pereboom, manipulation allows an actor to keep some elements of freedom and independence. For example, if someone manipulates me into joining a club by threatening to reveal something embarrassing about me, then I still have a choice as to whether or not I join. In this respect, event manipulated people maintain some element of independence. Similarly, Robert Kane argues that what is distinct about manipulation is that it influences our behaviour changing our preferences and desires, rather than influencing our actions directly (Kane 1999, p. 64). For Kane, there is at least some sense in which a manipulated person acts according to his or her own free will. With coercion, people act according to the will of another.

But there is more to manipulation than Pereboom and Kane's accounts. If manipulation just involved influencing behaviour by changing preferences then it would be no different from persuasion. Rather, manipulation alters our preferences through deception, or by taking advantage of a weakness.

Taking deception first, some argue that deception is a necessary feature of manipulation. For Robert Goodin, manipulation necessarily involves misdirection or fraud, because manipulation means that people are unaware that someone is changing their preferences (Goodin 1980, p. 9, 19). If we are aware that someone is trying to manipulate us, then our subsequent actions must come about through our own volition, so people can only be manipulated if they are unaware that it is happening. Of course, manipulation doesn't necessarily mean that that someone has intentionally misled us. I can manipulate someone if I know that they have an incorrect view but fail to rectify it.[4] But on this account manipulation does rely on some element of secrecy or deception.

But there is more to manipulation than just this. There are many situations where people are aware that someone is influencing their behaviour by changing their preferences, yet they are not acting according to their volition. In a series of progressive thought experiments, Joel Rudinow argues that deception isn't the only method through which manipulation operates (Rudinow 1978). Rather, manipulation also involves changing someone's behaviour by playing on a supposed weakness that they have. In either case, our behaviour is changed in a way that fails to respect our freedom to act as independent agents. People become objects of someone's will, even if they do not fully lose their independence. As such, manipulation is problematic for autonomy and it is incompatible with fair bargaining. This gives us three conditions for voluntariness. The first two conditions relate to information and rationality, whilst the third relates to manipulation.

---

[3]For discussion of manipulation, see: Kane 1999; Pereboom 2001; Wertheimer 1996.

[4]I this point from Christiano 1996, p. 118.

6.2.2 Coercion

Having argued that voluntariness requires three features (information, rationality and lack of manipulation) I now turn to a fourth. In addition to the requirements of bargaining ability and absence of manipulation, voluntariness also requires the absence of coercion. This is because coercion makes us subject to the will of another, thereby removing our ability to act as independent and autonomous agents. Coercion is also problematic for equality and justification in decisions. If someone's actions are subject to the will of another then they aren't able to advance their interests in a decision, nor are they able to justify their decisions to one another. Following our discussion from Chap. 4, fair bargains are therefore those that are free from unjustified coercion. I use the term: *unjustified* coercion because, as I go on to show, there may be some circumstances in which coercion is permissible in fair bargains.

But first it's necessary to consider what coercion entails. There is a large literature on coercion and it is a concept that is defined in a number of different ways. It's often thought that coercion involves getting someone to do something that they wouldn't have otherwise done. But this leaves no distinction between coercion and other forms of influence, such as manipulation and persuasion. As such, most authors argue that coercion involves forcing someone to do something. But there are very different understandings of what this 'force' involves.

For Robert Nozick, A coerces B if A communicates a credible threat to B which makes B change his or her actions (Nozick 1974). This means that coercion is based on threats and sanctions, rather than direct force. Coercion is problematic for voluntariness on this account because it removes an actor's autonomy. But whilst this meets some of our ideas about coercion, it also seems too narrow. Under Nozick's conception, coercive acts are limited to those that force an actor to do something through a threat. But not all coercive acts involve threats. As I argued in Chap. 4, the fact that an agent has no other reasonable alternatives means that its autonomy can be constrained regardless of whether there is a threat involved. The important point is that something is coercive if it prevents us from making a free choice (Olsaretti 2004, p. 119). This doesn't just happen if someone threatens us. It also happens if someone forces an agent to take an option, or if people aren't in a position to reject something. So it's also necessary to think about cases where there is a direct use of force.

Grant Lamond defines a coercive act as one that meets three conditions: (i) it forces an actor to do something against his or her will, (ii) it subjects the actor to the will of another, and (iii) the coerced actor is unable to do otherwise (Lamond 2000). This third condition is where Lamond's account diverges from Nozick's. It is not enough that an actor is threatened; an actor must also be forced to do something. This requires that there are no other reasonable alternatives available to the coerced actor. Coercion is important because it removes free choice; if a reasonable alternative is available, then the actor isn't forced into an agreement.

A 'reasonable' set of alternative options is a sufficient range of options so that an actor can remain autonomous. If an agent's options are sufficiently limited then they don't have a choice over whether or not they accept an agreement. Procedural justice means that each actor should be in a position that is sufficiently appealing so that they aren't forced into accepting an agreement, because they are in a position to reject the proposal. In the bargaining literature, this is sometimes referred to as a Best Alternative To a Negotiated Agreement (BATNA).[5] Bargainers with good BATNAs are in strong bargaining positions, whereas those with poor BATNAs are at a disadvantage. Procedural justice requires that bargainers have sufficiently good BATNAs so that they are not forced into accepting agreements. For Olsaretti, a sufficiently acceptable alternative is one that does not require us to give up any of our basic needs (Olsaretti 2004). This is one reasonable position to take on this issue. If people are in a situation in which they are unable to meet their most basic needs, then they are not in position to reject certain bargains, agreements or offers. If people are not in a position to reject something, then they have no choice and they are, in effect, forced into accepting the agreement.

But it's possible to force a person into a proposal even if they are able to meet their basic needs. I can coerce a wealthy person into accepting an agreement if I threaten to reveal incriminating information to the public about her. The wealthy person is able to meet her basic needs, yet she does not have a sufficient BATNA to reject my offer of blackmail: either she accepts the unfavourable agreement, or she faces defamation. Whilst basic needs provide one way for thinking about whether or not an actor has a suitable range of alternative offers, this is only one limited part of what is important. It's necessary to provide each actor with sufficient options so that each is in a position to reject an offer. Our concern is that there are some offers that people cannot refuse because they are in a position that is so unfavourable that they literally have no choice but to accept what is put in front of them, however unfavourable that offer actually is. Whilst Olsaretti's definition of a reasonable alternative option is appropriate to a certain extent, it's also important to think about when an option is so bad that actors have no choice but to accept a proposal.

I've argued that someone can be forced into accepting an offer when his or her alternative to the offer is sufficiently bad that the person has no choice but to accept the offer. I propose that there is no way to answer what a sufficiently bad option is without considering individual cases in their own context. That is, in order to think about whether an actor faces a sufficiently bad alternative range of alternative options, it's necessary to consider whether that actor is in a position to reject a proposal in individual cases. Olsaretti's definition of basic needs provides one way of doing this, but there is more at stake than just this. Procedural justice requires that each actor has a sufficient range of alternative options to an agreement, yet the only way to determine this is to think about whether an actor is in a position to reject an agreement in individual cases.

It's worth pointing out that simply having one option on the table does not mean that people do not have a reasonable BATNA. If someone's current situation is

[5]For discussion, see: Fisher et al. 1991.

sufficiently acceptable, then they are not forced into accepting an agreement, even if it is a 'take it or leave it' offer. Further, 'take it or leave it' offers are acceptable, provided that people are already able to meet their basic needs. It's also worth pointing out that if an actor has a very poor BATNA, then it does not follow that an actor is coerced in a bargain. If I'm offered a way out of an extremely unfavourable situation, then I hardly have a favourable BATNA. But it seems wrong to claim that I'm coerced on this occasion. What is important about having an insufficient BATNA is that actors are not in a position to reject offers that they would otherwise refuse. But this doesn't mean that having an insufficient BATNA necessarily implies that someone is coerced if they accept an agreement.

In addition to this account of coercion, some authors also emphasise the role of intention for coercion. For Lamond, it is not enough that a coerced actor is forced into doing something. Rather, if A coerces B, then B must be subject to A's will. This means that coercion involves deliberately imposing a disadvantage on another actor. For Lamond, the fact that one actor intentionally changes the will of another makes it particularly damaging for autonomy. But intention doesn't play a critical role here. People can be forced into acting in a certain way even if they are not directly subject to the will of another. Whilst intention may give us additional reasons for worrying about coercion, it is unclear why coercion doesn't involve acts that restrict our autonomy, even if it's unintended.

Alan Wertheimer offers an alternative account that avoids this problem by leaving intention aside. Wertheimer suggests that coercive arrangements involve two elements: (i) a choice element, whereby people are unable to exercise their free will, and (ii) a proposal element, whereby agreements take advantage of people (Wertheimer 1989, p. 32). The first element suggests that an act is coercive if an agent is unable to exercise his or her free will. The second element implies that an act is coercive if it takes advantage of someone. For Wertheimer, these are the necessary conditions for coercion. A coerces B if A creates a situation in which B has no choice to accept A's proposal if A acts wrongly in creating B's situation. This seems to capture our main concerns about coercion, whilst leaving aside the issue of intention. Coercion is important because of its implications for autonomy. But we are not so concerned with autonomy per se; we are concerned acts that affect our autonomy. The important point here is that coercive acts infringe an actor's independence and autonomy. In this respect, intention doesn't play a key role and Wertheimer's account seems more appropriate for our understanding of coercion.

But contrary to Wertheimer's account, coercion doesn't necessarily involve taking unfair advantage of anyone. Whilst situations in which people take unfair advantage on one another are important, this is a matter of exploitation, rather than coercion. As I show in the following section, exploitation is a separate matter, and coercion doesn't necessarily involve taking unfair advantage of someone. This means that we should accept the 'choice' element of Wertheimer's account, whilst leaving aside his 'proposal' element.

This leaves us with two necessary conditions for coercion: (i) there are no reasonable alternatives, and (ii) there is a credible threat or forceful action. Given that fair bargains are free from coercion, one necessary condition for fair bargains is that these conditions are not met simultaneously. Some authors suggest that this means that decision-makers should be substantively equal. The only way to avoid coercion, so the argument goes, is to make sure that actors cannot make use of favourable distributions of power and resources (Knight and Johnson 1997, p. 294). But substantive inequalities don't necessarily lead to coercive decision-making. Substantive equality is very difficult to achieve in multilateral institutions, and radical resource transfers are likely to be unfeasible in practice. Further, there is no reason for thinking that substantive equality is a necessary part of avoiding coercion. To be sure, substantive equality would remove the possibility of coercion here, but this doesn't mean that substantive symmetries between actors will lead to coercive decision. Rather, it's necessary to ensure that these asymmetries do not play a part influencing decisions.

Before moving on, it's worth considering whether coercion or manipulation are ever justified in fair bargains. As I said earlier, if we're considering bargaining among reasonable actors, then fair bargains are those that are free from coercion and manipulation. But, although there isn't sufficient room to give this question the full attention it deserves here, coercion and manipulation do seem justified if we're dealing with unreasonable actors. Unreasonable actors are those that are irrational, or who make extreme claims and proposals. These actors can hold up decision-making and obstruct agreement. Given the need to reach agreement on climate change, there may be some situations where coercion is justified in the UNFCCC. This might involve forcing unreasonable actors to accept an agreement or adopt more reasonable positions, or forcing uncooperative states to participate in, or comply with agreements. Coercion and manipulation can therefore play important roles in the UNFCCC.

6.3 Reciprocity and Exploitation

Thus far, I've argued that fair bargaining requires voluntariness and that this in turn requires sufficient information, rationality, and a lack of unjust manipulation and coercion. But this isn't all that's important here. There are some situations in which two parties enter an agreement voluntarily, each benefits, neither is coerced or manipulated, and yet people feel that there is something wrong with the agreement. This happens, for example, when bargainers receive highly disproportionate benefits from the agreement, or if the agreement leaves one party in an extremely disadvantageous position, leading us to think that the agreement is invalid because it is unconscionable in some way. I claim that such arrangements are unfair because they exploit one party. I suggest that fair bargains are those that allow actors to pursue their own self-interest to a limited extent, but that also involve some degree of reciprocity between actors. This is partly because parties should be able to advance their interests equally in a decision-making process.

I show this by drawing from the existing literature on exploitation and reciprocity. Starting with exploitation first, the literature generally suggests that an agreement is exploitative if one actor takes unfair advantage of another. Unfortunately, as many point out, this gives us as many different ideas about exploitation as there are about what fair treatment requires (Arneson 1992, p. 350). Several authors have tried to provide more concrete basis for exploitation, and the remainder of this section considers these claims, arguing that these are insufficient for thinking about exploitation.

6.3.1 Consent and Mutual Advantage

Some, such as David Miller, argue that exploitation depends on some defect in consent (Miller 1987). That is, an actor wouldn't freely enter into an exploitative agreement, so exploitation must conflict with our requirements of voluntariness set out above. On this account, exploitation only arises if an actor is coerced, or if the bargainers have incomplete information, or if some other condition of voluntariness is absent. If this is the case then it isn't necessary to consider the problem of exploitation at all. If fair bargains are those that meet the requirements of voluntariness, and if exploitation only arises when these are not met, then we've already set the conditions for avoiding exploitation.

But this seems a very limited account of exploitation. Whilst it's possible to imagine situations in which exploitation arises because an actor is coerced or manipulated, it's also possible to imagine situations in which people are exploited even if they fully consent to agreements.[6] If I forget my umbrella on a particularly rainy day, then I might think that I'm exploited if the only umbrella shop in town raises its prices to take advantage of my misfortune. But it seems strange to suggest that I don't consent to the agreement if I choose to buy an umbrella. I could brave the rain instead, and I am certainly not forced into the agreement. This suggests that exploitation doesn't depend on a defect in consent and a broader account of exploitation is needed than one that only focuses on this feature.

Given that I've already specified that fair bargains are those that are free from these issues, I'm particularly interested in cases where people feel that something is wrong even when an agreement is consensual. Further, given that actors only voluntarily enter agreements that are mutually beneficial, I'm concerned with consensual, mutually beneficial exploitation. Some authors, such as Joel Feinberg, say that exploitation cannot arise in consensual, mutually beneficial transactions. Feinberg claims that it is wrong to say that one party gains at another's expense if a transaction is mutually advantageous (compared to a no-transaction baseline), since both parties benefit from the arrangement (Feinberg 1990, p. 178). Given that people often think that exploitation involves taking advantage of someone, this

[6]For discussion, see: Wertheimer 1996, p. 247.

suggests that exploitation cannot occur in mutually beneficial transactions. I agree that exploitation can leave one party worse off than before. But, contrary to this view, I claim that it is too restrictive to say that an agent never gains at another's expense in mutually beneficial transactions. An agent might gain from another's expense by taking a highly disproportionate share of the benefit, even if both parties benefit from the transaction (Zwolinski 2007). This suggests that exploitation can occur in consensual, mutually advantageous transactions. It's therefore necessary to think about how exploitation can arise in such transactions.

6.3.2 Alternative Notions of Exploitation

I've argued that, contrary to Miller, exploitation isn't dependent on a defect in consent and, contrary to Feinberg, it needn't be mutually disadvantageous. But how else should we think about exploitative agreements? In what follows I suggest that our concerns about exploitation are most prevalent where people are left much worse off, or disadvantaged by a bargain. I argue that A exploits B when A takes advantage of a weakness of B in order to gain a highly disproportionate share of a bargain. In order to do this I first reject two common ways for thinking about exploitation. I then defend the claim that exploitation involves unfair advantage, and unfair outcomes.

Alan Buchanan offers one suggestion for thinking about exploitation, defining it as: 'the harmful, merely instrumental, utilisation…' of an individual '…for one's own advantage' (Buchanan 1985, p. 87). But whilst this captures some of our thoughts about exploitation, one might think that there are cases that aren't exploitative even if someone uses an individual for their own advantage, and people can benefit from others without exploiting them. If two gamblers sit down at a poker table then people don't think that the loser has been exploited when the winner walks away with the kitty. There are clearly cases where people can fairly benefit from a bargain, provided certain conditions are met.

Others suggest that it's important to focus more specifically on the fairness of the transaction itself. For example, when discussing the exploitation of sweatshop workers, Jeremy Snyder suggests that there is a sense in which people care about the fairness of the transaction, regardless of what the outcome is (Snyder 2008, p. 391). Snyder argues that a person exploits someone if they benefit from an unfair situation. This happens, for example, if a person is not in a position to reject a proposal. This seems a more reasonable position to endorse. People certainly think that consensual exploitation can arise if someone is in an unfavourable situation, otherwise the exploited actor wouldn't accept the agreement.

But the problem with this approach is that it isn't clear why an exploited actors needs to be in an unfair situation. It is possible to think of cases in which someone is exploited if they are in an unfavourable situation, regardless of whether this position is fair. Following our example from above, I'm not in an unfair position if I need an umbrella when it's raining, yet I am certainly in an unfavourable one. What's left is a notion of exploitation that relates to (i) an unfavourable position, and (ii) taking an unfair advantage. The first of these elements concerns the choices that are available to an actor, whereas the second concerns the distribution of the bargain (what Snyder calls 'a lack of reciprocation').

To this end, some theories of exploitation claim that there are at least two features at play: one procedural, and one substantive. For example, both David Miller, and Alan Wertheimer argue that an exploitative transaction is one that (i) arises from some advantage that one party has over another (the procedural element), and (ii) results in an outcome that is unfair for one of the actors (the substantive element) (Miller 1987; Wertheimer 1996). In what follows, I suggest that exploitation concerns both of these elements. I argue that the first part of exploitation concerns asymmetric bargaining power: one party uses an advantage to gain more from another in a transaction. The second part of exploitation concerns unfair advantage: exploitation occurs when one party uses this advantage to gain an unfair share of the gains from a transaction.

6.3.3 Exploitation (1): Asymmetric Bargaining Power

The first of these elements seems quite straightforward: A can only exploit B if A has a superior bargaining position (if we're considering consensual exploitation). If not, then it is difficult to see why B would accept the exploitative agreement. An actor is in a superior bargaining position if they have greater bargaining power. Although bargaining power is conceptualised in a number of different ways, it is often characterised as the relative ability of a party to align the outcome of a bargain with its own interests.[7]

Superior bargaining power can come about through different factors (Raiffa 1982, p. 54). Bargaining power can arise through the resources that an actor has, which are often categorised as either 'internal' or 'external'. Internal resources are those that are endogenous to the bargaining process itself, such as the ability to haggle, or the amount of information that an actor has about a decision. External bargaining resources are exogenous to the bargaining process, which includes issues such as the amount of material wealth that an actor has, or its political clout in the broader political context. Both internal and external resources influence the outcome of a bargaining process. An actor with more bargaining resources is able to shape the outcome of the bargain more than another. If an actor has greater bargaining power by virtue of either its internal, or external bargaining resources, then it is in a position to exploit another actor.

But bargaining power is also defined in terms of the relative value that each actor has on a negotiated agreement (Harstad 2010, p. 279). If actors place different valuations on reaching a mutually acceptable agreement, then one actor can reach a more favourable outcome by rejecting the agreement and holding out for a better

[7]See: Kverndokk 1995, p. 201; Zartman and Rubin 2000, p. 7; Weiler 2013.

outcome. If I possess a ticket to a concert that you are desperate to attend, then I can exploit you by virtue of my bargaining power. I know that you value the ticket more than I do and that you will go to great lengths to attend the concert. This is what some people refer to as 'hard' bargaining (in comparison to 'soft' bargaining) (Weiler 2013). One party can make threats or extract higher returns from an agreement because it values a mutual agreement less than another. Thomas Christiano says that this is how the US has traditionally acted in trade negotiations, where it has been able to negotiate advantageous agreements by making take it or leave it offers to those who are not in a position to refuse the agreement (Christiano 2009). Because developing countries are unable to turn these offers down, the US has been able to secure trade agreements that are highly asymmetrical.

This account of exploitation requires the absence of a range of reasonable alternative options. This raises familiar questions about what an adequate range of reasonable alternative options is. Earlier, I suggested that someone can be coerced in an agreement if they do not have a reasonable alternative option to the agreement. I then argued that there are situations where an agreement is exploitative even if it isn't coercive. But here it seems as though I am suggesting that an actor can be exploited if they are in an unfavourable bargaining position, where an unfavourable bargaining position can be defined in terms of the options that are available to an actor.

The problem is that this seems to support Miller's view (which I earlier rejected) that a defect in consent is necessary feature of exploitation. If an actor has no reasonable alternatives then he or she isn't in a position to refuse exploitative offers. According to this argument, a lack of reasonable alternative options is a necessary condition for exploitation. But this seems to pose a problem for my argument, which states that there can be voluntary, non-coercive exploitation. If a lack of reasonable alternatives is a necessary condition for both exploitation and coercion, then this argument no longer holds.

But this assumes that there is only a narrow understanding of what is meant by a 'reasonable set of alternative options'. Rather than describing whether or not someone faces a reasonable set of options as two distinct points, there might be a range of different positions. As Miller points out, this distinction is more likely to represent a spectrum with a range of different option sets rather than two binary positions (Miller 2009, p. 220, footnote 29). If there is one reasonable option available (including the option to reject an offer), then a person is coerced if he accepts that option. If someone faces some limited set of reasonable options, then he might not be coerced, but he can still be exploited. The point is that exploitation isn't dependent on a lack of reasonable alternative options. Rather, exploitation can arise in situations where people face some limited range of options, which means that rejecting the agreement is an extremely difficult thing to do. This is only a sufficient condition for exploitation. Asymmetries in bargaining power do not necessarily imply that one party is exploited. After all, if I offer to sell you the ticket at its face value, then I haven't exploited you. So there is more involved with exploitation than just asymmetric bargaining power.

6.3.4 Exploitation (2): Unfair Advantage

This is where the second element of exploitation comes into play. A transaction is exploitative if someone uses a favourable bargaining position to achieve a disproportionate share of the gains from a bargain. If 'asymmetric bargaining power' represents a procedural condition of exploitation, 'unfair advantage' represents a substantive condition. It is not just that parties have asymmetric bargaining power; exploitation means that one party uses this power to gain a disproportionate share of the gains from an agreement. Whilst selling the concert ticket at face value is not exploitative, demanding an exorbitant price clearly is. There is therefore a second necessary condition of exploitative transactions, which is that exploitative transactions involve taking unfair advantage of an actor.

But problems arise in thinking about what unfair advantage really means. In some cases, this is quite straightforward. If an actor is left in a very bad situation, or if there are highly disproportionate gains from an agreement, then it seems reasonable to say that one party has an unfair advantage. But then it is difficult to make more specific claims about what an unfair advantage is. In what follows, I consider some different options for thinking about this. For the most part, these all suggest that there should be some sort of equality in the arrangement, although there is a great deal of disagreement over what sort of equality this should be. Having done this, I then outline my own idea of how to think about disproportionate gains.

For Jeremy Snyder, exploitation is about failing to meet people's very basic needs. On this account, unfair advantage arises if someone is left in a situation where she is unable to meet her most basic needs. Snyder's subject is the exploitation of sweatshop workers, so his focus on basic needs is appropriate for his topic. But just because someone is left unable to meet her basic needs, it doesn't follow that she is necessarily exploited in a bargain. Imagine that a heavily indebted employee is discussing the terms of his employment with a new employer. The employer doesn't exploit the employee if he offers a very reasonable wage for the employee's services. This is true even if the employee is unable to provide for his basic needs, on account of his high debt. Snyder's account therefore seems a limited approach to exploitation and it's necessary think about cases where people are exploited even when their basic needs aren't compromised.

This leaves us in a difficult situation regarding unfair advantage. So far, I've suggested that fair bargains are those that are free from exploitation. I've also argued that there are two necessary conditions for exploitation: asymmetric bargaining power, and unfair advantage. But it is difficult to define exactly what unfair advantage entails. In response to this, I propose that, in situations where there are large asymmetries in bargaining power, fair bargains are those in which the final distribution meets a requirement of reciprocity. I outline and defend this claim below.

6.3.5 Reciprocity

My suggestion is that if actors have asymmetric bargaining power then fair bargaining requires that they bargain under a norm of reciprocity. This means that the more powerful actors make a 'fitting and proportional return for the good or ill we receive' (Becker 2005, p. 18). That is, reciprocity involves acting in the way that we would expect others to act if they shared our circumstances. People make the same concessions that they expect others to make and they make offers that they would accept themselves.

Rather than maximising their self-interest, actors should restrain their behaviour. This doesn't mean that people should offer equal terms, or that they should make altruistic proposals. But it does require some limits on our own self-interest when there are differences in bargaining power. This notion of restraint is something that Henry Shue has proposed for fair bargaining in climate negotiations. Shue argues that:

> Justice is about not squeezing people for everything one can get out of them... [A] commitment to justice includes a willingness to choose to accept less good terms than one could have achieved – to accept only agreements that are fair to others as well as to oneself. (Shue 1992, p. 385)[8]

In addition to Shue's argument, people should restrain their actions according to reciprocity. Not only should they constrain their self-interest when others are in an unfavourable position, but they should do so by offering terms that they would accept as reasonable were they in a similar situation. This stops people from taking unfair advantage of others. The problem with exploitation is not that someone benefits from an agreement, nor that someone benefits from a favourable bargaining position. Exploitation is problematic when someone benefits from a bargain in a *disproportionate* way through a particularly favourable bargaining position. If decision-makers have unequal bargaining positions then they should restrain their actions by treating others as they would think they should be treated if they were in the same situation.

It is possible to make several claims about what a fair distribution looks like in this respect. (i) The final distribution should not represent a highly asymmetric distribution of the gains from a bargain. This means that there should be at least some minimal degree of proportionality in the benefits that arise from an agreement. (ii) Neither party should be left in a worse situation than before the bargain. If both parties should gain some benefit from the transaction then neither party should be in a worse situation after the agreement. (iii) The outcome of a bargain should not worsen existing asymmetries in bargaining power. This is because exploitation does not just involve the distribution of the gains from cooperation. One party can exploit another by taking further advantage of an existing advantage in bargaining power. Fairness requires that bargains do not exacerbate existing asymmetries in

[8]Cecelia Albin makes a similar argument (Albin 2003, p. 377–8).

bargaining power. Taken together, these provide some minimal requirements for fair bargaining among actors with roughly equal bargaining power. These requirements are necessary because fair bargains are those that are free from exploitation. Whilst the exact requirements of exploitation are difficult to define, these criteria represent at least a starting for thinking about what's necessary to avoid exploitation in a bargain.

6.4 Issue Linkage

So far, I've argued that there are two conditions for fair bargains: (i) voluntariness, and (ii) reciprocity. This section builds on this by specifically considering the most prominent form of bargaining in the UNFCCC: issue linkage.[9] This involves 'linking' one policy issue to another, thereby allowing actors to exchange things that they value differently in order to reach a mutually acceptable agreement. Because this process involves trading concessions and gains, issue linkage is a case of a bargaining process.

Issue linkage is often classified as either positive, or negative. Positive issue linkage occurs when both parties gain from the agreement. In this sense, positive linkages are 'integrative bargains', which are those for which the total payoff available to actors increases through the process of bargaining (Raiffa 1982, p. 144). This happens, for example, if two actors value something differently, or if each party has a production advantage over another. In relation to the UNFCCC, positive linkages might concern linking issues such as trade and emissions mitigation. For example, some suggest that Russia's acceptance of the Kyoto Protocol was a direct condition of its accession to the WTO.[10] The idea here is that Russia was initially reluctant to join the Kyoto Protocol. Yet Russia valued participation in the WTO very highly, whereas many members of the WTO valued Russia's participation in the Kyoto Protocol. Given that these two groups valued Russia's participation in these two agreements, they were able to reach a mutually acceptable agreement by finding a mutually acceptable agreement.[11]

In contrast to positive linkage, where both parties gain from the agreement, negative linkage involves one party being made worse off than before. Given that bargainers will not voluntarily enter agreements that make them worse off, negative linkage is associated with the use of coercion, either through sanctions or threats.[12] Barrett and Stavins set out several negative linkages that states make in the

---

[9] See: Barrett and Stavins 2003.

[10] See, for example: Chasek et al. 2006, p. 126; Hall et al. 2010, p. 672; Karp and Zhao 2010; Andonova and Alexieva 2012.

[11] For discussions of how climate change might be linked to trade, see: Stokke 2004, p. 350; Ghosh and Woods 2009, p. 473; Harstad 2009, p. 292.

[12] For a discussion of threats in decisions, see: Knight and Johnson 1997, p. 294.

UNFCCC, including reciprocal measures, financial penalties and trade restrictions (Barrett and Stavins 2003, p. 363). The authors also note that the threat of trade restrictions played an important role in ensuring the participation and compliance of a large number of states in the Montreal Protocol. The key point here is that, in contrast to positive linkage, negative linkage involves making an actor worse off than before, or threatening to make an actor worse off than a pre-bargain baseline.

6.4.1 Related and Unrelated Issue Linkage

In addition to the distinction between positive and negative linkage, there is also a distinction between linking issues that are 'related' or 'unrelated'. Related issues are those that have some sort of connection, or overlap in some way. This means that taking action on one policy area directly affects another. There are several examples of this happening in climate institutions, and two of the most prominent are the Clean Development Mechanism (CDM) and Reducing Emissions for Deforestation and Forest Degradation (REDD). The CDM is a mitigation scheme that links emission reduction projects in developing countries to emissions trading schemes in industrialised countries. The idea is that this promotes development whilst also reducing carbon emissions. REDD connects forestry projects to emissions mitigation policies, thereby realising joint gains between these two ends. Both of these policies involve linking different areas (development and mitigation, and forestry and mitigation) in order to realise joint gains. But it's also possible to link wholly unrelated issues in multilateral institutions. These are issues that aren't directly related: changing policies in one issue area doesn't affect the other. For example, an unrelated issue linkage might involve linking extradition treaties to trade agreements. Alternatively, territorial disputes could be linked to environmental matters.

6.4.2 Issue Linkage and Procedural Efficiency

Some suggest that issue linkage is important for mutually acceptable multilateral agreements. Issue linkage increases the potential for parties to reach agreement on contested decisions. For this reason, many decision theorists advocate incorporating a wide range of issues and dimensions in multilateral negotiations.[13] This is something that has been taken up in the literature on environmental governance, which includes discussions about the potential benefits of linking environmental

[13]For example: Raiffa 1982; Susskind and Ozawa 1992, p. 152; Susskind 1994, p. 87; Harstad 2010, p. 291.

issues to: energy security,[14] economic development,[15] financial side-payments,[16] technology transfer,[17] and trade.[18]

But some argue that issue linkage is not always favourable from the point of view of procedural efficiency. Cecelia Albin criticises issue linkage on the grounds that it complicates negotiations that are already technically dense (Albin 2003, p. 14 footnote).[19] Climate change negotiations are already difficult to manage and introducing more issues is likely to put additional burdens on decision-making processes that are already strained. Limiting negotiations to a single issue keeps things simple. Other authors suggest that issue linkage may lead to obstructionism. Andresen and Wettestad argue that increasing the number of issues at stake in a negotiation makes it more likely that potential controversies and 'blocking points' will arise (Andresen and Wettestad 1992). This might be particularly relevant for the UNFCCC, where many issues are politically sensitive. Attempting to link sensitive issues to other policy areas rather than dealing with them in their own right may stir up greater resistance to progress and agreement. Whilst supporting issue linkage for some reasons, James Sebenius suggests that linking divisive issues may destroy the possibility for an agreement on otherwise tractable issues in multilateral negotiations (Sebenius 1984, p. 182–217).

This shows that, whilst it is sometimes desirable for making decisions easier, issue linkage also has its downsides. There are strong arguments on both sides of the debate of whether issue linkage improves the possibility of reaching agreement in the UNFCCC, which boil down to a matter for empirical analysis beyond the scope of this book. Whilst it isn't possible to make any concrete statements about the desirability of issue linkage here, we can make some remarks about how the UNFCCC might respond in light of this debate. One solution is to take both sides of this argument into account when making proposals about what issues decision-makers can bring to the table. We can acknowledge that increasing issue linkage can improve the chance of reaching a successful agreement, whilst also remaining wary that issue linkage can complicate and exacerbate already difficult negotiations.

In this chapter, I've done two things. First, I've argued that there are two conditions for fair bargains: (i) voluntariness, and (ii) reciprocity. Voluntariness means that bargainers are informed and that bargains are free from manipulation and coercion. Reciprocity means that each decision-maker puts restraints on the sorts of demands that it can make from other actors. Provided that these conditions are met, procedural justice allows actors to pursue their own self-interest in bargains.

---

[14]Hall et al. 2010, p. 672.

[15]Depledge and Yamin 2009, p. 448.

[16]Barrett 2001; Barrett and Stavins 2003; Sheeran 2006.

[17]Carraro and Siniscalco 1998.

[18]Barrett 1995, 1997.

[19]See also: Susskind and Ozawa 1992; Mitchell and Keilbach 2001, p. 894 footnote.

Second, I've also shown how one particular form of bargaining, issue linkage, takes place in multilateral institutions. The following section combines these strands in order to make some policy recommendations for issue linkage in the UNFCCC.

6.5 Issue Linkage in the UNFCCC

Given that fair bargaining allows actors to make decisions on issues that they would otherwise disagree in a fair way, issue linkage is a desirable feature of COP negotiations. At the same time, it's necessary to be cautious about issue linkage in the UNFCCC on two accounts. On the one hand, it's important to recognise that increasing issue linkage can over complicate negotiations, or where it introduces issues that are divisive. COP negotiations are already extremely complicated and face many procedural difficulties. Expanding the number of issues at stake in these negotiations can cause more problems than it solves. On the other hand, there are clearly some limits that we want to put on bargaining for the sake of fairness. In what follows, I draw on the normative arguments from sections two and three to stipulate what types of linkages are permissible in the UNFCCC.

But before doing this, it's worth saying something about how these constraints might be enforced. In the UNFCCC, this could be done through the Chair of the COP. The Chair could act as an objective third party actor who oversees the negotiation process and monitor issue linkages and bargains between states. Bargains would have to meet this actor's approval, according to set procedural criteria that promote fair bargains. Decision-makers, including both member states and NSAs, could also be able to raise concerns about manipulation and coercion with the Chair in order to prevent specific sorts of linkage. Having made some suggestions about how this might be done, it's now possible to say something about the sort of constraints that should be put on issue linkage in COP negotiations.

6.5.1 Constraint (1): Negative Linkage

I've argued that fair bargains are those that are free from unjustified manipulation and coercion. This is because fair bargains are those that voluntary, and these are two necessary conditions for voluntariness. For this reason, unjustified negative issue linkage should be prohibited in UNFCCC decisions. Decision-makers should not be allowed to make unjustified threats to one another, nor should they be allowed to force each other into agreements because these sorts of linkages go against the conditions of voluntariness set out earlier.

But this doesn't mean that the UNFCCC should prohibit negative issue linkage altogether. As I suggested in section two, there may be some cases for which both coercion and manipulation are justified in fair bargains. This might happen if decision-makers are unreasonable in the sense that they intentionally obstruct agreement, or make extreme demands. In these cases it might be necessary and permissible to use threats and sanctions in order to encourage these actors to act reasonably. The question of whether or not states can legitimately impose threats or sanctions on those who are not part of a climate change agreement is too great to take up here. However, the point that I want to make is that negative issue linkage cannot be ruled out as a possible policy response to unreasonable actors. Although these measures should be used with caution. Whilst threats and sanctions may be permissible if actors are unreasonable, using these measures is likely to be very detrimental to the overall negotiation process. This might lead to hostility within the institution and a breakdown of cooperation. Whilst manipulation and coercion may be important parts of a climate agreement that successfully addresses climate change, there may also be severe downsides to this approach.

6.5.2 Constraint (2): Manipulation, Coercion and Exploitation

Further to negative issue linkage, there are other conditions for fair issue linkage in COP negotiations. The UNFCCC should ensure that issue linkage does not bring about manipulation, coercion or exploitation. Earlier, I suggested that people shouldn't take advantage of those who are not in a position to reject an offer. I also argued that reciprocity also means that the terms of the agreement should not represent a highly asymmetric distribution of the benefits of cooperation that the outcome of a bargain should not perpetuate existing disadvantages. This means that, where states have insufficient BATNAs, the UNFCCC should prevent other states from taking advantage of them. This doesn't mean that it should prohibit linkages with these actors. This would be doubly unfair, since it would prevent those who are already worse off from enjoying the benefits of issue linkage. . Just because manipulation, coercion and exploitation *can* arise if an actor has an insufficient BATNA, it does not mean that these issues *will* arise. Rather than preventing bargaining between those that don't have sufficient BATNAs, the UNFCCC should act to ensure that bargaining processes that involve these actors are fair, despite the fact that some actors do not have sufficiently fair alternatives to a negotiated outcome. But we do want to ensure that these actors aren't manipulated, coerced, or exploited in the bargaining process by virtue of the fact that they are not in a position to turn refuse an offer.

Further, the UNFCCC should prevent bargains that worsen existing power asymmetries. This might mean restricting linkages to issue areas outside of climate change. For example, Burtraw and Toman argue that there are several potential areas in multilateral negotiations where there may be the potential for coercive use of unrelated issues (Burtraw and Toman 1993, p. 128). These include, for instance, the Jackson-Vanik Amendment (1974), which was a provision of US federal law that denied the "most favored nation" status to trading partners that constrained free emigration and other human rights. Likewise, Cecilia Albin also argues that it is unethical to offer economic aid to poor states in exchange for entering agreements that compromise their safety (Albin 2003, p. 14 footnote). The UNFCCC should prevent states from linking issues where there is a possibility that states may be able to exploit advantageous positions in other issue areas.

6.5.3 Constraint (3) Side-Payments

Further to these conditions, the UNFCCC should also discourage the use of side-payments in COP negotiations. Whilst positive issue linkage links two policy areas to each other, side-payments specifically involve the transfer of material wealth for the purpose of making an agreement more attractive to a decision-maker.[20] Some argue that side-payments play an important role in creating incentives for participation and compliance in multilateral institutions.[21] But linking issues to monetary values is also problematic for decisions, and there are at least two reasons why side-payments between states should be discouraged.

The first problem is that using side-payments introduce moral hazard. That is, if states see that others are receiving compensation for their endorsement of an agreement then they may make similar demands in exchange for their own endorsement. Alternatively, providing side-payments for cooperation may lead to a shift in attitudes away from supporting public institutions for collective action and towards the expectation of a payment. Of course, this is also likely to arise with other forms of issue linkage. If a state links one policy to another to encourage state participation, then other states may demand the same benefits. But there is something distinct about monetary transactions that specifically encourages perverse incentives.

A second problem is that side-payments are sometimes seen as a form of bribery. This is problematic because these sorts of linkages are seen as unfair or disrespectful. Bribery carries many negative connotations and is something that's been brought up within the UNFCCC. At COP15 some states saw the offer of adaptation funds in return for their agreement as a bribe.[22] The delegation of Cuba was most vocal on this issue, claiming that it represented 'blackmail'. The point here is that there are at least some linkages that are damaging for an institution more generally, aside from considerations about what it means for specific bargaining parties. Given that they involve the direct transfer of wealth, side-payments seem to present more problems in terms of manipulation, coercion and exploitation. For this reason, the UNFCCC should limit the use of side-payments in bargains that take place in the COP.

[20]For discussion in relation to the trade regime, see: Ghosh 2010, p. 3. For a discussion in the context of climate change: Burtraw and Toman 1993, p. 131.

[21]DeSombre 2002, p. 19; Barrett and Stavins 2003; Viguier 2004, p. 197.

[22]For discussion, see: Dimitrov 2010, p. 813.

But this does not rule out all monetary transfers between states. Some transfers, such as those relating to compensation for environmental damage or development aid for adaptation, may be important as a matter of justice. These sorts of payments differ from side-payments to the extent that they are based on what is owed to people as a matter of principle, as opposed to improving the desirability of an agreement for self-interested bargainers. So whilst the UNFCCC should limit the use of side-payments, this does rule out all monetary transfers between states. The relevant standard for a justified payment is whether it is based on principle, rather than self-interest.

6.6 Conclusion

Bargaining is an important part of the UNFCCC. I've argued that fair bargains are those that are voluntary and reciprocal, which means that bargains are free from manipulation and coercion, and that they take place among informed and rational actors. Following these points, there are several constraints that the UNFCCC should put on the sorts of linkages that are made in COP negotiations. But this represents just one answer to the question of how to resolve difference in the face of disagreement. In the next chapter, I turn to a second response: voting.

References

Albin, C. 2001. *Justice and fairness in international negotiation*. Cambridge: Cambridge University Press.

Albin, C. 2003. Getting to fairness: Negotiations over global public goods. In *Providing public goods: Managing globalization*, ed. I. Kaul, P. Conceição, K. Le Goulven, and R.U. Mendoza. Oxford: Oxford University Press.

Andonova, L.B., and A. Alexieva. 2012. Continuity and change in Russia's climate negotiations position and strategy. *Climate Policy* 12(5): 614–629.

Andresen, S., and J. Wettestad. 1992. International resource cooperation and the greenhouse problem. *Global Environmental Change, Human and Policy Dimensions* 2(4): 277–291.

Arneson, R. 1992. Exploitation. In *Encyclopedia of ethics*, ed. Lawrence C. Becker, 350–352. New York: Garland.

Barrett, S. 1995. *Trade restrictions in international environmental agreements*. London: London Business School.

Barrett, S. 1997. The strategy of trade sanctions in international environmental agreements. *Resources and Energy Economics* 19(4): 345–361.

Barrett, S. 2001. International cooperation for sale. *European Economic Review* 45(10): 1835–1850.

Barrett, S., and R.N. Stavins. 2003. Increasing participation and compliance in international climate change agreement. *International Environmental Agreements: Politics, Law and Economics* 3: 349–376.

Becker, L.C. 2005. Reciprocity, justice, and disability. *Ethics* 116(1): 9–39.

Buchanan, A. 1985 *Ethics, efficiency, and the market*. Totowa: Rowman and Allanheld.

Burtraw, D., and M.A. Toman. 1993. Equity and international agreements for CO2 constraint. *Journal of Energy Engineering* 118(2): 122–135.
Carraro, C., and D. Siniscalco. 1998. International environmental agreements: Incentives and political economy. *European Economic Review* 42: 561–572.
Chasek, P.S., D.L. Downie, et al. 2006. *Global environmental politics*. Boulder: Westview Press.
Christiano, T. 1996. *The rule of the many: Fundamental issues in democratic theory*. Boulder/Oxford: Westview Press.
Christiano, T. 2009. Democratic legitimacy and international institutions. In *The philosophy of international law*, ed. S. Besson and J. Tastioulas. Oxford: Oxford University Press.
Depledge, J. 2008. Striving for no: Saudi Arabia in the climate change regime. *Global Environmental Politics* 8(4): 9–35.
Depledge, J., and F. Yamin. 2009. The global climate change regime: A defence. In *The economics and politics of climate change*, ed. D. Helm and C. Hepburn. Oxford: Oxford University Press.
DeSombre, E.R. 2002. *The global environment and world politics*. London: Continuum.
Dimitrov, R.S. 2010. Inside UN climate change negotiations: The Copenhagen conference. *Review of Policy Research* 27(6): 795–821.
Feinberg, J. 1990. *The moral limits of the criminal law, Volume 4: Harmless wrongdoing*. New York: Oxford University Press.
Fisher, R., W. Ury, et al. 1991. *Getting to yes: Negotiating agreement without giving in*. New York: Penguin.
Ghosh, A. 2010. *Making climate look like trade? Questions on incentives, flexibility and credibility*, Policy brief for centre for policy research. New Delhi: Dharma Marg.
Ghosh, A., and N. Woods. 2009. Governing climate change: Lessons from other governance regimes. In *The economics and politics of climate change*, ed. D. Helm and C. Hepburn. Oxford: Oxford University Press.
Goodin, R.E. 1980. *Manipulatory politics*. New Haven/London: Yale University Press.
Hall, D., M. Levi, et al. 2010. Policies for developing country engagement. In *Post-Kyoto international climate policy: Implementing architectures for agreement*, ed. J.E. Aldy and R.N. Stavins. Cambridge: Cambridge University Press.
Harstad, B. 2009. *The dynamics of climate agreements*. CMS-EMS Discussion Paper 1474. http://www.kellogg.northwestern.edu/faculty/harstad/htm/nt.pdf.
Harstad, B. 2010. How to negotiate and update climate agreements. In *Post-Kyoto international climate policy: Implementing architectures for agreement*, ed. J.E. Aldy and R.N. Stavins. Cambridge: Cambridge University Press.
Kane, R. 1999. *The significance of free will*. New York: Oxford University Press.
Karp, L., and J. Zhao. 2010. Kyoto's successor. In *Post-Kyoto international climate policy: Implementing architectures for agreement*, ed. J.E. Aldy and R.N. Stavins. Cambridge: Cambridge University Press.
Knight, J., and J. Johnson. 1997. What sort of political equality does democratic deliberation require. In *Deliberative democracy*, ed. J. Bohman and W. Rehg. Cambridge, MA: MIT Press.
Kverndokk, S. 1995. *Tradeable CO2 emission permits: Initial distribution as a justice problem*. CSERGE GEG Working Paper, 92–35.
Lamond, G. 2000. The coerciveness of law. *Oxford Journal of Legal Studies* 20(1): 39–62.
Miller, D. 1987. Exploitation in the market. In *Modern Theories of Exploitation*, ed. A. Reeve. London: Sage.
Miller, D. 2009. Democracy's domain. *Philosophy & Public Affairs* 37(3): 201–228.
Mitchell, R.B., and P.M. Keilbach. 2001. Situation structure and institutional design: Reciprocity, coercion, and exchange. *International Organisation* 55(4): 891–917.
Nash, J. 1950. The bargaining problem. *Econometrica* 28: 155–152.
Nozick, R. 1974. *Anarchy, state and Utopia*. New York: Basic Books.
Olsaretti, S. 2004. *Liberty, desert and the market: A philosophical study*. Cambridge: Cambridge University Press.
Pereboom, D. 2001. *Living without free will*. Cambridge/New York: Cambridge University Press.
Raiffa, H. 1982. *The art and science of negotiation*. Cambridge, MA: Harvard University Press.

Rudinow, J. 1978. Manipulation. *Ethics* 88: 338–347.

Sebenius, J.K. 1984. *Negotiating the law of the sea*. Cambridge, MA: Harvard University Press.

Sheeran, K.A. 2006. Side-payments or exemptions: The efficient climate control. *Eastern Economic Journal* 32(2): 515–532.

Shue, H. 1992. The unavoidability of justice. In *The international politics of the environment: Actors, interests and institutions*, ed. A. Hurrell and B. Kingsbury. Oxford: Clarendon Press.

Snyder, J.C. 2008. Needs exploitation. *Ethical Theory and Moral Practice* 11(4): 389–405.

Stokke, O.S. 2004. Trade measures and climate compliance: Institutional interplay between WTO and the Marrakesh accords. *International Environment Agreements: Politics, Law and Economics* 4: 339–357.

Susskind, L. 1994. *Environmental diplomacy: Negotiating more effective global agreements*. New York/Oxford: Oxford University Press.

Susskind, L., and C. Ozawa. 1992. Negotiating more effective international environmental agreements. In *The international politics of the environment*, ed. B. Kingsbury and A. Hurrell. Oxford: Clarendon Press.

Viguier, L. 2004. A proposal to increase developing country participation in international climate policy. *Environmental Science & Policy* 7: 195–204.

Weiler, F. 2013. Determinants of bargaining success in the climate change negotiations. *Climate Policy* 12(5): 552–574.

Wertheimer, A. 1989. *Coercion*. Princeton: Princeton University Press.

Wertheimer, A. 1996. *Exploitation*. Princeton: Princeton University Press.

Zartman, I.W., and J.Z. Rubin. 2000. *Power and negotiation*. Ann Arbour: University of Michigan Press.

Zartman, I., and S. Touval. 2010. *International cooperation: The extents and limits of multilateralism*. New York/Cambridge: Cambridge University Press.

Zwolinski, M. 2007. Sweatshops, choice, and exploitation. *Business Ethics Quarterly* 17(4): 689–727.

Chapter 7
Voting

7.1 Introduction

The UNFCCC requires some sort of voting mechanisms to make collective decisions. The Rules of Procedure of the COP state that, in cases where parties cannot reach agreement by consensus, decisions should be made by majority rule (UNFCCC RoP Rule 42). But these rules have not been formally adopted, meaning that the UNFCCC makes decisions by consensus and decisions are only adopted if none of the decision-makers openly objects to the decision. The problem with a decision-making rule that depends on universal support to reach an outcome is that individual actors have the power to block agreements. This stifles ambition and leads to outcomes that reflect the 'lowest common denominator' (Prins and Rayner 2007, p. 974). One way around this is to adopt a majority rule voting procedure, which is increasingly seen as a potential solution to some of the procedural problems of the UNFCCC.[1]

Despite the support for using an alternative voting mechanism to the consensus rule, little has been said about the fairness of such rules in relation to the UNFCCC, or about the voting weights that should be used if this option was pursued. Voting weights determine the amount of influence, or say that each actor has in a decision. The UNFCCC currently specifies that each actor should have one vote in a decision (UNFCCC 1992 Article 18). This isn't an issue if decisions are made by consensus, since every decision-maker has an equal opportunity to block a decision by objecting to it. Giving more votes to an actor therefore has no impact on the final decision. But voting weights are important if decisions are made by majority rule. Multilateral institutions that use majority rule for making decisions assign voting weights in very different ways, and this is something that has important implications for the overall fairness of the decision. Some institutions, such as the UN, use a system of voting

[1]For support of this view see: Harstad 2009, p. 3; Rajamani 2011a; Biermann et al. 2012.

L. Tomlinson, *Procedural Justice in the United Nations Framework Convention on Climate Change*, DOI 10.1007/978-3-319-17184-5_7

parity by applying a rule of 'one state-one vote', which emphasises the importance of the state.[2] Other institutions, such as the IMF and the World Bank, assign votes in a way that reflects the financial contribution that each actor makes to the institution. Given that there are great disparities in the population size of different countries, there are also questions about what role population should play fair voting processes.

This means that there are two issues at stake in the design of fair voting process. First, there are questions about the choice of the voting procedure that the UNFCCC should use. Second, there are questions about how the UNFCCC should weight votes in its decision. This chapter uses the normative criteria introduced in Chap. 5 to consider both of these issues in turn. In doing so, it makes three policy recommendations for voting in the UNFCCC.

(1) Decisions should be made by majority rule, but only after considered deliberation
(2) Votes should be weighted in part according to the number of individuals that each state represents
(3) In specific circumstances, states that are wholly unrepresentative of their constituencies should be excluded from decisions

7.2 Voting Rules

Voting rules differ in respect to the proportion of votes required for a decision to be formally adopted.[3] Generally speaking, there is a spectrum of voting rules that range from unanimity at one end, to simple majority rule at the other. Fair voting rules are those that promote the criteria for procedural justice set out in Chap. 4: autonomy, the equal advancement of interests, and justification. The epistemic quality of a voting rule, and the extent to which it promotes reasonableness are also important issues.

But voting procedures are important for other values too. For one thing, the choice of voting procedure is important for procedural efficiency, or the ability to reach a decision quickly. The choice of voting procedure (and voting weight) also determines the say that each actor has in a decision-making process, which might have important implications for whether or not an actor chooses to participate in a multilateral agreement. Whilst fairness is valuable in its own right, it is also important not to exclude other considerations before making specific policy recommendations for voting reform. This section introduces the most frequently used voting rules in multilateral institutions: unanimity, consensus, and majoritarian

[2]UNFCCC RoP Rule 41, paragraphs 1–2. The Rules of Procedure use the term 'regional economic integration organizations' rather than 'coalition'.

[3]For discussions of voting in multilateral institutions, see: McIntyre 1954; Blake and Payton 2009; Payton 2010.

voting.[4] It considers each of these forms of decision-making in order to determine what procedural fairness requires, ultimately arguing that majority rule is important for fairness and efficiency.

7.2.1 Unanimity

Unanimity effectively gives each decision-maker an equal vote in the decision-making process, since any one member can either accept the decision or veto it by abstaining from the vote.[5] This seems important for procedural justice for a number of reasons.

Because it depends on finding an outcome that is mutually acceptable to all, unanimity ensures that no actor has to accept a policy that it does not endorse as correct. This means that unanimity does not infringe independence or autonomy by forcing a person to go against his or her will. In this respect, unanimity means that decision-makers act as authors of the decisions that they are subject to, thereby allowing them to act as autonomous agents in decisions.

Unanimity encourages actors to deliberate and justify their reasons to one another. If each actor has a veto right over a decision, then everyone's position has to be recognised and accommodated in the decision-making process. This means that decision-makers listen to each other, exchange information, and acknowledge each other's opinions. It also helps actors to form opinions that acknowledge the views and positions of others. In this respect, unanimity serves an important function in encouraging actors to justify their decisions to one another.

Unanimity also seems fair because it gives each actor an *equal* say over a final decision, rather than privileging any single actor. Unanimity therefore promotes equality in at least one sense, to the extent that it gives each participant an equal degree of influence over the final outcome.

But giving each an equal power to veto a decision is also detrimental to equality in a different sense. By giving each actor the right to veto a decision it, unanimity privileges those who favour the status quo.[6] Each member of a group has the power to prevent a new option from being collectively adopted. Under unanimity, one actor can veto a decision, even if there is consensus from the rest of the group. Contrary to giving equal consideration to each actor's interests, which was one of the requirements of procedural justice that I introduced in Chap. 5, this gives greater say to those who favour the status quo. This is the primary reason for thinking that unanimity is inappropriate for fairness.

Unanimity also rewards those who hold extreme views. Because it's necessary to get everyone on board in order to make a decision, actors who do not value an

[4]For more on these rules, see: Steinberg 2002; Touval 2010, p. 83.

[5]For discussion of this principle, see: Zamora 1980; Blake and Payton 2009; Touval 2010, p. 83.

[6]For discussion, see: Christiano 1996, p. 88.

agreement as much as others can hold out for what they want rather than compromising on an issue. In doing so, unanimity can discourage people from deliberating with one another and encourage them to adopt hard bargaining strategies with the promise of getting what they want. Jorgen Wettestad suggests that this is something that happened in the early stages of the negotiations for the UNFCCC, where major oil producers advocated unanimity so that they could later block emission reduction commitments (Wettestad 1999, p. 216).

Aside from concerns about fairness, there are also other problems with unanimity. For one thing, voting procedures should provide a way of reaching agreement even when people are unable to arrive at a unanimous agreement. As I argued in Chap. 2, there is reasonable disagreement over some issues in the UNFCCC. If this is the case, unanimity either leads to stalemate or brings about outcomes that represent the lowest common denominator (Sohn 1974, p. 445; Zamora 1980, p. 571). Several authors suggest that the use of unanimity in the UNFCCC has held up decisions and stifled ambition (Harstad 2009, p. 3; Dimitrov 2010; Payton 2010, p. 1).

7.2.2 *Consensus*

If unanimity is problematic for fair decision-making, then an alternative voting rule to consider is consensus. Recent discussions show that the meaning of consensus is open to interpretation and still subject to much academic debate.[7] Whilst consensus is traditionally understood as the absence of expressed objection, Lavanya Rajamani suggests that this definition should be expanded to include cases where there is some opposition to a decision (Rajamani 2011a, p. 515). This is something that's become evident in the COP itself. For example, Jacob Werksman notes that in several other cases the COP has adopted decisions even though some parties have objected to the agreement (Werksman 1999, p. 12). This was also evident at the UNFCCC COP16 in Cancun (2010) where the Chair accepted the negotiation text despite express opposition to the decision by Bolivia. These represent more lenient standards for accepting an agreement than unanimity, since they can accommodate some limited disagreement.

By appealing to a less demanding standard for agreement, consensus avoids at least some of the deadlock and political stalemate associated with unanimity. A decision can be adopted without the endorsement of every actor, so long as there isn't any expressed objection or abstention. Consensus therefore represents a compromise between the need to push ahead with a decision and the need to respect each actor's position on an issue. This is a significant benefit in situations where there is a premium on making decisions quickly.

[7] For example: LRI 2011; Rajamani 2011b; for the consensus rule in multilateral institutions, see: Van Houtven 2002; Steinberg 2002; Rajamani 2011a, p. 516.

Despite these benefits, consensus fails to address the problems that led us to reject unanimity. Like unanimity, consensus still faces the problem of stalemate when there is reasonable disagreement on something. It also fails to meet our requirement for the equal consideration of interests, since it gives those who favour the status quo the ability to prevent agreement. A further concern is that consensus masks disagreement rather than resolving it (Brunnée 2002, p. 10). If a group presses ahead with a decision even though some parties harbor disagreement, then this might just postpone conflict until some later time. For example, some saw the COP20 negotiations in Lima as simply delaying discussions around the most contentious issues until COP21 in Paris the following year.[8] Whilst consensus can provide a more efficient way of making decisions, it might mean that some actors later renege on the outcome of the decision because they feel that their interests weren't sufficiently taken into account during the decision.

In practice, the search for consensus can be just as elusive as unanimity and may lead to political stagnation. It's also unclear why consensus is an improvement in terms of fairness. Given that the UNFCCC is renowned for its lack of progress, and given the need for urgent action on climate change, it is worth considering the merits of majority rule procedures as a possible remedy.

7.2.3 *Majority Rule*

Some now see majority rule as a desirable alternative to consensus and unanimity in multilateral institutions.[9] Some might think that majority rule doesn't fare well in terms of autonomy because it forces people to accept decisions that they do not endorse. It is true that some element of independence is lost in a majority rule process. But majority rule doesn't mean that people give up all of their independence. For one thing, people still express their interests and views in the deliberative processes. People also maintain an element of autonomy through the voting process itself, expressing their interests or judgements about decisions. Whilst unanimity may score better on the grounds of autonomy in some sense, it's also important to consider how to make decisions when people disagree.

Some might also criticise majority rule in terms of deliberation. The purpose of deliberation is to arrive at a mutual consensus. In this respect, one might argue that majority rule is inappropriate because it can bypass deliberative discussion, when decision-makers should really be trying to reach mutual agreement (Sohn 1974, p. 441). But this is an incorrect view of the role of majority rule in deliberative discussion. Majority rule doesn't replace deliberation, nor is it a procedure that people should turn to once decision-makers have failed to reach a consensus. Rather, majority rule is a way of expressing our judgement about something

[8]Michael Levi makes this point: Levi 2014.

[9]For example: Biermann *et al.* 2012; for discussion: Blake and Payton 2009, p. 5; Brunnée 2002.

after deliberative discussion. Jeremy Waldron convincingly argues that aiming for consensus in deliberation doesn't mean that it should be supported as the appropriate political outcome (Waldron 1999, Chap. 5). Rather, people should first deliberate and then make a decision by a fair voting rule. In this sense majority rule is fully compatible with deliberation.

Whilst the desirability of majority rule is questionable in terms of autonomy or deliberation, there are at least two ways in which it is important for fairness. Importantly, majority rule is more favourable than either unanimity or consensus on some accounts of equality. Earlier, I suggested that there is some sense in which unanimity is an unequal voting rule, because it gives undue power to a minority. Majority rule, on the other hand, is fairer in this respect. Under majority rule, it is possible to give decision-makers different numbers of votes and in some situations, this is important for equality. Assuming that actors are roughly equal, in the sense that they are roughly affected by a decision to the same extent, then giving each actor an equal vote in majority rule voting procedure is fair because it gives each an equal say in that decision (Christiano 1996, p. 88). But actors are unlikely to be equal in this respect for many decisions. As I argue later, there are many cases where it is important to give more votes to certain actors, if people have different stakes in a decision, or if a decision-maker represents more actors. This isn't possible under unanimity or consensus, since under these rules, each decision-maker has the same power to either support or reject an agreement. For this reason, majority rule can be more advantageous for the equal consideration of interests, because, unlike consensus, it is possible to give actors different amounts of votes, which is important for fairness in some cases.

A second way that majority rule is important for fairness concerns respect. For Waldron, majority rule respects individuals by accommodating some difference over different conceptions of justice and the common good (Waldron 1999, p. 111). Striving for consensus can play down difference, encouraging us to think that someone is wrong, or incorrect if they disagree with us. But majority rule allows individuals to maintain their difference. It accommodates the fact that people might simply disagree on some matters and in this sense it respects individuals where unanimity and consensus do not.

These two arguments give us strong reasons for thinking that majority rule is more desirable than unanimity and consensus when it comes to fairness. There're also good reasons to think it's advantageous in terms of procedural efficiency. But there's still a mixed picture here and there are at least two potential objections to majority rule that need to be addressed.

One might object that majority rule is not a fair way of making decisions since it allows a minority to rule over a decision. Buchanan and Tullock argue that those who hold more orthodox views support this claim (Buchanan and Tullock 1962, p. 243). The orthodox argument is that if people have a choice between two options, A and B, and more than a simple majority is needed to reach a decision (i.e. more than 51 % of the vote) then it's possible to end up with a situation in which the minority controls the final decision. That is, a majority of 75 % is needed for a decision and 74 % support A whilst 26 % support B, then the minority can prevent the adoption of option A. According to this argument, the minority controls whether or not a decision is made. Buchanan and Tullock rightly criticise this view, arguing that it equates positive decisions to authorise group action with negative decisions that block actions proposed by others. Whilst the minority can prevent the adoption of option A, it cannot bring about the adoption of option B. To this extent, neither group controls the outcome, but rather blocks an action from coming about.

A second concern is that majority rule marginalises those who are in the persistent minority (Bodansky 1999, p. 607). If a minority has preferences that are consistently different to the majority, then that group is always overruled in the decision-making process. This might mean that an entire group is marginalised and disenfranchised from collective decisions. This is a serious concern, but there are ways of resolving this problem and it doesn't mean that majority rule should be rejected altogether. Individuals should be required to listen to one another and take into account each other's interests during deliberation, which should encourage the majority to take the interests of the minority into account when making decisions. In extreme situations, where a minority is significantly affected, it might be necessary to grant veto rights to a minority group for certain decisions. This might involve appointing independent actors to observe decisions and determine whether a certain group is persistently marginalised by decisions to a significant degree. These are measures that should be taken in light of specific decision-making contexts, and the point here is to give some preliminary thoughts about how these problems can be avoided.

So majority rule is not just a fair way of making decisions; it also scores better in terms of other important values. Further, I've shown that it stands up to two potential objections. But this is only one side of the coin, and the following section takes up the second challenge of how to weight votes in a fair way.

7.3 Vote Weighting

Having determined that majority rule is a fair way of making decisions, this section now turns to the fairness of vote weighting. The allocation of voting weights determines how much influence each actor has in a decision-making process and this varies significantly between different multilateral institutions.[10] In this section, I look at some of the prominent proposals for how votes should be weighted in decisions. I consider whether votes should be weighted according to: (i) the stake that an actor has in a decision, (ii) how many people it represents, and (iii) how accurately it represents them. I argue that votes should be weighted according to these rules provided that certain criteria are met.

[10]For vote weighting in multilateral organisations, see: Blake and Payton 2009, p. 5–4.

7.3.1 Proportionality

Let's consider (i) first – the magnitude of someone's stake in a decision. It is often claimed that fair, or democratic voting processes are those that give each voter an equal vote (Kelsen 1955; Saunders 2010). The idea is that each member of a group should have an equal say over collective decisions. But some authors suggest that fairness requires that those who have a greater stake in a decision receive more votes. This is known as the proportionality principle, and it implies that those who are affected more by a decision should have a greater say in how it is made.[11] In what follows, I introduce the argument for proportionality and discuss its merit as a principle of fairness. I argue that it is a fair way of making decisions *when actors are voting to advance their interests in a decision.* I then go on to argue that proportionality is not a fair way of making decisions *when voters are voting to express their judgement or opinion in a decision.*

I start by looking at Brighouse and Fleurbaey's comprehensive review of the proportionality principle, in which they give four arguments for favouring it over equal suffrage (Brighouse and Fleurbaey 2010, p. 138). Their first argument is based on premise that fair decisions are those that treat individuals with equal respect. For the authors, equal respect means giving equal consideration to each individual's interests in a decision. In this respect, Brighouse and Fleurbaey share Thomas Christiano's justification for democracy (Christiano 1996, 2008, p. 78). But the problem with Christiano's approach is that giving each actor an equal say in a decision doesn't give equal consideration to each actor's interests when some are affected by a decision more than others. For Christiano, this means that a necessary condition for democracy is that every member of a group is affected by its decisions to roughly the same extent, which puts a strong restriction on the sorts of decisions people can make. For Brighouse and Fleurbaey, however, if some are affected by a decision more than others, then these actors should be given more votes for the sake of achieving an equal consideration of interests.

This is the main argument for proportionality. But there are other arguments in support of this principle. The proportionality principle is also relevant for fairness because it is important for autonomy. If individuals have unequal stakes in a decision, then giving each an equal vote means that those who are more affected by a decision are subject to the will of those who are less concerned. By giving each actor a say according to the stake that they have, the proportionality principle gives individuals greater control over these decisions, thereby enhancing autonomy and independence.

Proportionality also appeals to epistemic concerns. Those who are more affected by a decision are more likely to be better placed to make decisions about it.

[11] For discussion, see: Arrhenius 2005; Tännsjo 2005; for support of this principle, see: Warren 2002; Brighouse and Fleurbaey 2010.

This means that giving more votes to the more affected should lead to better outcomes, in the sense that these outcomes have higher epistemic quality. Although this isn't an argument for fairness, it does give additional support to proportionality.

There are also instrumental reasons for supporting this principle. Brighouse and Fleurbaey argue that proportionality can avoid the problem of persistent minorities. If decisions are made through a simple majority, then some actors may be persistently disenfranchised from a decision-making process. But proportionality can help alleviate this problem by giving a greater say to those who are affected by a decision more than others. A minority that is significantly affected receives a greater say, preventing situations where a majority persistently overrules one particular group.

Brighouse and Fleurbaey claim that these arguments give us good reason for adopting the proportionality principle. The authors acknowledge that there are likely to be practical problems of implementing the policy and recommend that it is used as a guideline rather than as a basis for developing a mathematical formula for vote weighting. Bell and Rowe give an example of how this could be done in local climate policy in the UK. The authors endorse proportionality whilst drawing quite general policy recommendations, arguing that those who are worse off in society are likely to be the most affected by its decisions and that greater voting power should be given to those in this group (Bell and Rowe 2012). This seems a sensible step to take. Determining how much each actor is affected by a decision and how this should translate into voting power is going to be subject to disagreement, if not practically demanding. But this doesn't prohibit us from making some broad proposals for implementing proportionality in practice.

The idea that proportionality promotes autonomy and epistemic value whilst avoiding disenfranchisement seems plausible enough. But the crux of Brighouse and Fleurbaey's argument lies in its appeal for the equal consideration of interests, which is more controversial. If people make decisions for the sake of advancing their own interests, then Brighouse and Fleurbaey's argument for proportionality is appropriate for procedural justice. But people don't just make decisions for the sake of promoting their own interests, and the proportionality no longer seems so appropriate when people vote to express a judgement or opinion about something. The important point for this argument lies in the distinction between judgements and interests. Christiano argues that whereas an interest is something that is an important component of wellbeing, a judgement is a belief about a fact (Christiano 1996, p. 54). A judgement about something can either be correct or incorrect, whilst an interest cannot.

Recalling our discussion of the burdens of judgement from Chap. 2, people might disagree because they are mistaken, or because they hold different world views, or because of subjective bias. Given that disagreement arises not just from mistake, but also from difference, people should respect each actor's judgement in a decision, even if they think that it is wrong. This doesn't mean that people should give equal weight, or consideration to each person's judgement. If one person is an expert on an issue then people might give that person's opinion greater attention when they are making up their own minds about something. But when people to express their judgement in a voting process, each should have an equal say in the decision. This is the only way of giving each actor sufficient respect, given the fact of reasonable disagreement brought about through the burdens of judgement.

This means that proportionality is not a fair way of making decisions *when voters are voting to express their judgement or opinion in a decision*. Whether the principle of proportionality is an appropriate principle of procedural justice depends on the sort of decision that voters are being asked to vote on.

7.3.2 Proportional Representation

Having discussed whether voting rights should depend on the extent to which an agent is affected, I now ask whether multilateral voting rights should depend on the population size that each voter represents. In multilateral institutions, state representatives vote on behalf of the domestic constituencies that they represent. Some institutions, such as the UN General Assembly and the UNFCCC, operate on a policy of 'one state, one vote', which is often referred to as 'sovereign equality' (McNicoll 1999, p. 411; Rajamani 2006). This prioritises the state above other types of actor, giving each state the same vote regardless of its population, leading some to claim that it is undemocratic (Nye 2001; Müller et al. 2003, p. 4–7). Other institutions assign votes according to the population that each state has.[12] It's necessary to think about what fairness requires when voters represent different numbers and different types of actors. In what follows, I claim that votes should be distributed according to the number of actors that each voter represents.

Fair decision-making processes are those that allow individuals to advance their interests equally in a decision-making process. This means that each individual's interests should be valued to the same extent as any other's. Where voters act as representatives, procedural justice means that are those who are represented have their interests advanced equally in a decision. What matters here is the equal advancement of the interests of individuals represented in a decision, rather than the interests of those who are actually voting. If voters vote to advance the interests of those that they act on behalf of, then each voter should receive a number of votes proportionate to the number of actors that it represents. If voters act as representatives in a decision, then there is a strong case for thinking that each voter should have a say according to the number of actors that they represent.

[12] McIntyre 1954, p. 494; Müller et al. 2003, p. 4–7; Biermann et al. 2012.

7.3.3 *Democratic Legitimacy*

Earlier I identified three issues concerning representation. Having discussed two of them, I now turn to the third. If votes should be weighted according to the number of actors that each voter represents, then it is also important to think about how representative each voter is of those that they claim to represent. In this section, I claim that those decision-makers that are poor representatives should not receive fewer votes by virtue of their representativeness. Those decision-makers that are completely unrepresentative should, however, receive fewer votes.

In the last section I argued that voters should have more votes if they represent more actors. But decision-makers can be poor representatives, in the sense that they fail to effectively translate the interests and views of their constituents in decisions. In the UNFCCC, many state delegations act on behalf of autocratic governments that can hardly be said to provide fair representation to those that they act on behalf of. Giving a voter who represents more actors more votes makes little sense if that voter is a poor representative on those of whom it is supposed to act on behalf. For this reason, one might think that proportional representation means that fewer votes should be given to those that are poor or inefficient representatives.

But there are problems with this approach. For one thing, if a voting actor is a poor representative of its constituency, but it is still nonetheless the best representative available, then giving that voter fewer votes is doubly unfair. Not only is that constituency poorly represented, but it has less of a say as well. It therefore seems unfair to deprive people of any representation in decisions by virtue of the fact that they are poorly represented. This means that actors that are poor representatives of their constituencies shouldn't receive fewer votes in a decision, even though some of these actors do not represent their constituents accurately. Whilst it isn't ideal that some states are poorly representative, states are still the best representatives at the multilateral level, and voting rights shouldn't be limited on this basis.

But the situation changes if actors are wholly unrepresentative of those that they claim to act on behalf of. Whilst there might be some justification for refraining from giving poor representatives fewer voting rights on the basis that these are the nonetheless the best actors for the job, this isn't the case if a decision-maker is completely unrepresentative of its constituency. If procedural justice means that actors should be given a certain number of votes on account of the constituency that it represents, then it is pointless to give an unrepresentative actor votes if it completely fails to act on behalf of its constituency, because the actor isn't fulfilling the role that warrants giving it more votes in the first place.

These three principles give us an idea about what fairness requires in terms of vote weighting. Votes should be weighted in a way that takes into account the stake that an actor has in a decision, but only if voters are making decisions that advance the interests of their constituents. Further, votes should be weighted according to the number of actors that each voter represents, and whether or not an actor is representative of its constituency. Before considering how this should be interpreted in the UNFCCC, it's worth considering two other common methods of weighting votes.

7.3.4 Contribution

One such method is to distribute votes according to the contribution that each actor provides to a cooperative arrangement. This is what happens in the IMF and World Bank, where votes are weighted according to the financial contribution that each member state makes to the institutions.[13] Some claim that this is done for the sake of historical precedent and political feasibility as much as it is for fairness.[14] Still, it's worth considering whether or not this is a fair way of weighting votes.

One thing to note about contribution straightaway is that there are a number of ways that it can be interpreted. In the case of climate change, contribution might be thought of in terms of emissions reductions (potential or actual), financial terms (both in relation to mitigation *and* adaptation), or some combination of these. It's not clear why a particular interpretation of contribution is important for procedural justice.

Taking contribution more generally, there might be some cases where it is appropriate to distribute voting rights in this way. People often think that shareholders should have a say in a company's decision that is proportional to their shareholding. Likewise if several states collaborate in the development of an infrastructure project, then it seems reasonable to suggest that states should have a say in proportion to their contribution to the project.

Whilst contribution is an appropriate way of making decisions in some situations, it is not appropriate for fair decisions in the UNFCCC. Contribution seems most appropriate when (i) there is a clearly defined good for contribution, (ii) where it is important to give parties incentives to contribute more, and (iii) where decision-makers are primarily affected by the extent to which they contribute to the decision. In these sorts of cases, each party benefits from the contribution principle, because each decision-maker has an incentive to contribute more. Further, if the amount each actor contributes largely determines the extent to which actors are affected then the contribution principle seems to appeal to a notion of proportionality, rather than contribution to a collective arrangement. The contribution principle also seems most appropriate when decision-making power should be determined by the resources that each actor has. In some cases, giving decision-makers power according to their contribution is important, because these are the actors who are primarily affected by these decisions, and it is important that these actors are encouraged to contribute more to the cooperative arrangement.

This is not the case with the UNFCCC, where decisions have profound implications for people on a global scale, regardless of their contribution to the institution (however it is interpreted). Given the vast disparities in state resources, and given the high stakes involved in climate change, the principle of contribution is an

[13]For discussion of voting in the World Bank, see: Blake and Payton 2009, p. 5–4; For discussion of voting in the IMF, see: Van Houtven 2002, p. 5; Rapkin and Strand 2005; Payton 2010, p. 3–4.

[14]Several authors argue that these voting arrangements arose due to feasibility issues during the design of the institution: Lister 1984, p. 37; Rapkin and Strand 2005, p. 1994; Payton 2010, p. 2.

inappropriate way of weighting votes. The contribution principle would allow wealthier states, or those with high emissions profiles, to dominate decisions that affect people around the world. Whilst contribution seems an appropriate principle in some cases, it is not suitable for the context here.

7.3.5 Moral and Technical Competence

In addition to contribution, some argue that those who are better at making decisions should receive more votes. According to Brighouse and Fleurbaey, there are at least two respects in which some people are more competent at making decisions (Brighouse and Fleurbaey 2010). People might be more technically competent, if they have the skills or knowledge to make good decisions. People can also be morally competent, in the sense that they are better at understanding different viewpoints and acting on behalf of others.

Taking technical competence first, many have considered whether it's desirable to give more votes to those who are better placed to make good decisions. I argued in Chap. 2 that fairness requires that decision-makers act under a standard of reasonableness, which includes some minimal standards of rationality. Here, it's worth thinking about whether decision-makers should receive votes on the basis of technical competence. As I suggested in Sect. 7.3.1, those who are more affected by a decision may be better placed to make good judgements about an issue. Alternatively, it might be that those who are better educated or more rational are better at making judgements. Arriving at a correct outcome means taking into account the soundness of each individual's judgement. For John Stuart Mill, people rightly feel offended if they are ignored in a decision, but this doesn't mean that people can't attach greater weight to those who are experts (Mill 1861, Chap. 8). This means that people should respect each actor but only according to his or her expertise or ability. Jeremy Waldron also argues that people could endorse some plural voting scheme based on rationality, whilst acknowledging that there are likely to be many practical problems in doing so (Waldron 1999, p. 115).

But this is only true if all that is important is arriving at a correct outcome. Here, fairness is important too. There is also an important difference between giving greater consideration to the views of experts and giving those experts a greater say in the decision-making process. Brighouse and Fleurbaey argue that influence in a deliberative stage amounts to providing information and evidence about a decision (Brighouse and Fleurbaey 2010, p. 10). But giving greater consideration to the views of experts in the deliberative stage does not amount to giving some more votes than others. Rather, fairness requires that each actor receives equal respect in a decision-making process, which involves giving each an equal say (provided that they are equal in the terms discussed above).

But there are other reasons for thinking that technical competence is not so relevant for vote weighting. Just because someone is a technical expert, it doesn't mean that he or she will make decisions that represent the interests of others. As such, giving that person more votes goes against giving equal consideration to each person's interest. Weighting votes according to technical competence also raises concerns about autonomy. Giving more votes to those who are better able to make decisions gives people less control over decisions, reducing their independence, which is also problematic for fairness. People also disagree on some issues because they interpret the world differently. This follows our discussion of the burdens of judgement from Chap. 2. Given that people disagree because they have different worldviews and experiences (rather than just because of mistake and self-interest) fairness means that people should respect each actor's judgement equally, even if some are better at making decisions than others. This means that procedural justice does not require giving people votes on the basis of technical competence, provided that decision-makers meet some minimal standards of rationality.

Turning to moral competence, Brighouse and Fleurbaey argue that some people are less morally competent than others. In fact, the authors argue that people often think that trustees should be delegated with power when people are unable to meet some minimal level of moral competence. This happens when people make decisions on behalf of children. This implies that people should only be given votes if they meet some minimal requirements of moral rationality. But this doesn't mean that those who are more morally competent should receive more votes in a decision. Rather, votes should be given to people who are sufficiently rational, and then give each an equal say thereafter. But, as I argued in Chap. 2, fair decisions are those that involve reasonable actors, where reasonable actors are those who meet some minimal standards of rationality. So it seems reasonable to assume that moral competence should not play a role in determining how votes are weighted in the UNFCCC.

To conclude this section: neither moral nor technical competence appear relevant factors that we should take into account when thinking about how to weight votes fairly, and fairness requires that that votes are weighted according to proportionality, representation, and representativeness.

7.4 Voting in the UNFCCC (1): Majority Rule

Having identified what procedural justice requires in terms of voting, this section now considers how these requirements might be best implemented in the UNFCCC. Foremost, in line with the arguments above, the COP should make decisions by majority rule, but only after sufficient reasoned deliberation. Sufficient reasoned deliberation requires making sure all viewpoints are heard and taken into consideration in a debate, where a viewpoint concerns an opinion or position on a matter. It also means that each actor makes an effort to understand and acknowledge the views and opinions of others, rather than voting without hearing all of the issues at stake.

As I argued in Sect. 7.2, majority rule gives a fair say to each actor, rather than privileging those who favour the status quo. It also avoids many of the procedural problems associated with unanimity and consensus. In this respect, majority rule is more favourable both in terms of fairness, and procedural efficiency. But deliberation also offers some protection to those who are in a persistent minority, by encouraging all decision-makers to take into account each other's interests and act on behalf of others. Whilst this is not an outright solution to the problem of persistent minorities, it should at least provide one way of orientating decision-makers towards the common good and encouraging people to acknowledge the interests of others when they come to vote.

This measure also ensures that majority rule is not merely a substitute for deliberative discussion, but rather reflects an extension of the deliberative procedure. Fair decision-making processes require majority rule. But fairness also requires that decision-makers deliberate with one another. This is one of the fundamental features of procedural justice. But the epistemic value of deliberation is also important, helping each decision-maker to understand the views and interests of others. The UNFCCC can maintain these benefits by ensuring that there is sufficient deliberation and discussion of all views before taking a decision by majority rule.

These voting procedures should require a high proportion of votes in order for a decision to be adopted. This means that a supermajority is required to accept a decision, rather than a simple majority. This goes some way towards preventing the disenfranchisement of a minority group. Requiring a high proportion of votes makes it less likely that one particular group is persistently disenfranchised in a decision. It also encourages deliberation and discussion, because those who form the majority have to convince a larger group of actors of the merits of their position.

This approach also offers a partial solution to problems of noncompliance. Part of the reason that unanimity is so desirable in multilateral institutions is that states are only expected to comply with rules that they consent to. This makes it unlikely that a state will refuse to comply with the rules that result from cooperative agreement. But under majority rule situations might arise where a minority strongly disagrees with the outcome of a decision endorsed by a majority. Given that multilateral institutions depend on voluntary compliance, majority rule might either lead some states to exit the agreement, or fewer states to sign up in the first place. A high majority mitigates this problem to some extent. If a high majority is needed to make a decision then fewer actors will be left disgruntled by the outcome of a decision that they do not support.

There is at least one further reason for thinking that this objection is not overly troubling. For one thing, this objection assumes that the commitments that states make under consensus are the same as those that would arise from majority rule. Consensus encourages compliance, but it also means that any outcome reflects the lowest common denominator. In COP negotiations, this has often led to outcomes that are unambitious, or intentionally ambiguous. Under majority rule, some actors might not endorse the outcome, even if it comes about from a fair procedure. But the outcome of the decision should be more ambitious (given that they do not reflect the lowest common denominator). Whilst there may be greater defection from an agreement under majority rule, it is important to remember that this is a stricter rule overall.

7.5 Voting in the UNFCCC (2): Equal Time to Views in Debates

The second measure for promoting fairness in the voting processes of the UNFCCC is that, within a deliberative forum, each viewpoint should be given the same time and attention for discussion, irrespective of the population size of each voter. This means giving each perspective on an issue the same time for debate, rather than each actor's individual view, or each actor, the same time for presentation in debate. The primary reason for doing this is to ensure that each view and opinion on a matter is considered in a decision-making process. What's important for fairness is that each view is heard, rather than each view being repeated by all those who share it. Whilst people should take into account the fact that more people share a particular view or judgement than others, this doesn't mean that they should give more time or say according to how many people share that view.

There is an epistemic benefit to taking this approach. Giving each viewpoint time in a debate means that all views are actually heard, which should make it more likely that decision-makers will make correct judgements when they come to vote (where people vote on issues that are either correct or incorrect). A supporting argument is that giving each view equal time in a debate respects each actor's views, ensuring that each actor knows that his or her interests are presented and heard in a debate. This is important for fairness, but it is also important for other issues such as trust and legitimacy. As I argued in Chap. 3, it isn't just important that the UNFCCC *is fair*; people should *perceive* this institution as fair. Given this, there is additional merit to giving each viewpoint the same amount of time in debates.

7.6 Voting in the UNFCCC (3): Population Weighting

A third measure is that votes should be weighted, at least in part, according to population size. At the same time, it should be recognised that states are also important actors in their own right. To see this point, it's worth considering some of the differences in population size between different states. As of 2010, the UN estimates the population of China at 1,341,335,000 whilst the population of Tuvalu is 10,000.[15] Implementing a policy of 'one state, one vote', means that a state that represents over a billion individuals has the same decision-making power as one that represents several thousand.

Flipping the scenario, by making voting rights directly proportionate to population size, means that China has voting power equivalent to 100,000 times that of many other states. In fact, the populations of certain states, including China and India, are so large that distributing voting rights in this way would mean that these

[15] World Population Prospects 2010.

countries dominate almost every decision within the UNFCCC. This would not only leave many actors entirely disenfranchised from decisions, it would also means that we would ignore the importance of states in their own right. Whilst fairness requires that votes are weighted according to population size, states are also important actors beyond any representative role that they play.

It's necessary to find some middle ground here. We should respect the fact that different states represent vastly different numbers of individuals in COP negotiations whilst recognising that states represent group interests and collective identities, which are important matters aside from the size of their population. This already happens in some political systems. For example the political system of the US has one body that represents each state equally (the Senate) and one body representing states according to population size (the House of Representatives).[16] Similar measures could be taken so that voting weights are distributed in a way that takes both of these issues into account. One way of doing this is to assign each actor a set number of votes regardless of its population size (at the most basic level, this means one vote each). Additional votes could then be distributed according to the population that each state represents. For example, this might mean one additional vote per 10 million-population size. In the event that there are least some states that represent extremely small populations, each actor could receive some minimum number of votes so that none is put at a severe disadvantage in a voting process. This gives some attention to the size of the population that each state represents, whilst also respecting each state actor in the decision-making process.

7.7 Voting in the UNFCCC (4): Excluding Unrepresentative States

Finally, the UNFCCC should question the presence of dictatorial regimes and autocratic governments that are entirely unrepresentative of their domestic constituents. Whilst fairness requires that votes are weighted according to population size, there are clearly some states that are wholly unrepresentative of their constituents and the UNFCCC should avoid giving votes to these states. That said this, several things should be kept in mind.

Foremost, there is a difference between states that are wholly unrepresentative and those that are simply poor, or inaccurate representatives. There are many forms of representation in world politics and not all of them depend on democracy.[17] States shouldn't be excluded on the basis that they are not democratically accountable. Whilst the UNFCCC should avoid giving votes to wholly unrepresentative states, this doesn't mean that fewer votes should be given to poor representatives. Many member states of the UNFCCC are undemocratic, and even those that are

[16]For discussion, see Balinski and Young 2001, Chap. 2.

[17]For discussion, see: Kelsen 1955, p. 7.

democratic do not always represent the interests of their constituencies accurately. The important criterion here is whether a state is a worse representative of its constituent than other bodies would be. If a state is so unrepresentative that it is possible to identify a different actor that would be better at representing the people of a country then the UNFCCC should exclude states on the grounds that they are unrepresentative. At the same time, it should recognise that states are still often the best representatives of their domestic populations and, where this is the case, these states shouldn't be excluded, even if they are poor representatives.

A further issue is that it might be important for the UNFCCC to continue to engage with these actors as part of a process of encouraging them to democratise. Some people argue that engaging with autocratic states in multilateral institutions can play an important role in this respect (Adesnik and McFaul 2006; Levitsky and Way 2006). The idea is that engaging with autocratic actors in these contexts is more likely to promote norms of democracy and lead to better outcomes overall than isolating these states from global affairs. In this respect, the UNFCCC might continue to engage with these actors as part of an on going process to promote democratic practices. At the same time, these sorts of policies require continuing efforts to promote democratic norms across the multilateral political spectrum.

7.8 Conclusion

Reforming the voting procedures of the UNFCCC is one way of making decisions both fairer and more efficient. Certainly, many now consider majority rule an important option for resolving deadlock in the UNFCCC. But little has been said about the fairness of these measures, or about the choice of voting weights if these policies are adopted. This chapter has shed some light on this matter, by arguing that majority rule, subject to some constraints, is a fair method of decision-making, and that the UNFCCC should pursue population weighting for voting rights.

References

Adesnik, D., and M. McFaul. 2006. Engaging autocratic allies to promote democracy. *The Washington Quarterly* 29(2): 7–26.

Arrhenius, G. 2005. The boundary problem in democratic theory. In *Democracy unbound*, ed. F. Tersman. Stockholm: Stockholm University.

Balinski, M.L., and H.P. Young. 2001. *Fair representation: Meeting the ideal of one man, one vote*. Washington, DC: Brookings Institute.

Bell, D., and F. Rowe. 2012. *Are climate policies fairly made?* York: Joseph Rowntree Foundation.

Biermann, F., K. Abbott, et al. 2012. Navigating the anthropocene: Improving earth system governance. *Science* 16.335(6074): 1306–1307.

Blake, D.J., and A.L. Payton. 2009. *Decision making in international organizations: An interest based approach to voting rule selection*. Research in International Politics Workshop, January 16, 2008.

Bodansky, D. 1999. Legitimacy of international governance: A coming challenge for international environmental law? *The American Journal of International Law* 93(3): 596–624.

Brighouse, H., and M. Fleurbaey. 2010. Democracy and proportionality. *Journal of Political Philosophy* 18(2): 137–155.

Brunnée, J. 2002. COPing with consent: Law-making under multilateral environmental agreements. *Leiden Journal of International Law* 15(1): 1–52.

Buchanan, J.M., and G. Tullock. 1962. *The calculus of consent: Logical foundations of constitutional democracy*. Ann Arbor: University of Michigan Press.

Christiano, T. 1996. *The rule of the many: Fundamental issues in democratic theory*. Boulder: Westview Press.

Christiano, T. 2008. *The constitution of equality: Democratic authority and its limits*. Oxford: Oxford University Press.

Dimitrov, R.S. 2010. Inside Copenhagen: The state of climate governance. *Global Environmental Politics* 10(2): 18–24.

Harstad, B. (2009). Rules for negotiating and updating climate treaties. *Harvard Project on International Climate Agreements*. Belfer Center for Science and International Affairs. John F. Kennedy School of Government.

Kelsen, H. 1955. Foundations of democracy. *Ethics* 66(1): 1–101.

Levi, M. 2014. The lima climate agreement. *Council for Foreign Relations blog post*, 15th December 2014. http://blogs.cfr.org/levi/2014/12/15/the-lima-climate-agreement-isnt-as-new-as-it-seems/

Levitsky, S., and L.A. Way. 2006. Linkage versus Leverage. Rethinking the international dimension of regime change. *Comparative Politics* 38(4): 379–400.

Lister, F. 1984. Decision-making strategies for international organizations: The IMF model. *Series in World Affairs* 40(4).

LRI. 2011. *A guide to the UNFCCC institutions*, Legal response initiative briefing paper. London: Legal Response Initiative.

McIntyre, E. 1954. Weighted voting in international organizations. *International Organization* 8(4): 484–497.

McNicoll, G. 1999. Population weights in the international order. *Population and Development Review* 25(3): 411–442.

Mill, J.S. 1861. *Considerations of representative government*, On liberty and other essays. Oxford: Oxford University Press.

Müller, B., J. Drexhage, et al. 2003. Framing future commitments: A pilot study on the evolution of the UNFCCC greenhouse gas mitigation regime. *Oxford Institute for Energy Studies* EV 32.

Nye, J.S. 2001. Globalization's democratic deficit: How to make international institutions more accountable. *Foreign Affairs* 80(4): 2–6.

Payton, A.L. 2010. *Consensus procedures in international organizations*. EUI Working Papers 22.

Prins, G., and S. Rayner. 2007. Time to ditch Kyoto. *Nature* 449: 973–975.

Rajamani, L. 2006. *Differential treatment in international environmental law*. Oxford: Oxford University Press.

Rajamani, L. 2011a. The Cancun climate change agreements: Reading the text, subtext and tea leaves. *International & Comparative Law Quarterly* 60(2): 499–519.

Rajamani, L. 2011b. The climate regime in evolution: The disagreements that survive the Cancun agreements. *Climate and Carbon Law Review* 136.

Rapkin, D.P., and J.R. Strand. 2005. Developing country representation and governance of the international monetary fund. *World Development* 33(12): 1993–2011.

Saunders, B. 2010. Democracy, political equality, and majority rule. *Ethics* 121: 148–177.

Sohn, L.B. 1974. Introduction: United Nations decision-making: Confrontation or consensus. *Harvard International Law Journal* 15(3): 438–445.

Steinberg, R.H. 2002. In the shadow of law or power? Consensus-based bargaining and outcomes in the GATT/WTO. *International Organization* 56(2): 339–374.

Tännsjo, T. 2005. Future people, the all affected principle, and the limits of the aggregation model of democracy. In *Democracy unbound*, ed. F. Tersman. Stockholm: Stockholm University.

Touval, S. 2010. Negotiated cooperation and its alternatives. In *International cooperation: the extents and limits of multilateralism*, ed. I. Zartman and S. Touval. New York/Cambridge: Cambridge University Press.

UNFCCC. 1992. *United Nations framework convention on climate change*. Convention Text.

Van Houtven, L. 2002. *Governance of the IMF: Decision-making, institutional oversight, transparency, and accountability*. Washington, DC: International Monetary Fund.

Waldron, J. 1999. *Law and disagreement*. Oxford/New York: Oxford University Press.

Warren, M.E. 2002. What can democratic participation mean today? *Political Theory* 30(5): 677–701.

Werksman, J. 1999. Procedural and institutional aspects of the emerging climate change regime: Do improvised procedures lead to impoverished rules? *Concluding Workshop for the Project to Enhance Policy-Making Capacity Under the Framework Convention on Climate Change and The Kyoto Protocol*. London: Foundation for International Environmental Law and Development.

Wettestad, J. 1999. *Designing effective environmental regimes: The key conditions*. Cheltenham: Edward Elgar.

World Population Prospects. 2010. *World population prospects: The 2010 revision*. Population Division of the Department of Economic and Social Affairs of the United Nations Secretariat. http://esa.un.org/unpd/wpp/index.htm.

Zamora, S. 1980. Voting in international economic organizations. *American Journal of International Law* 74: 588–599.

Chapter 8
The UNFCCC: A Necessary Ideal

8.1 Introduction

Chapter 2 showed that there is reasonable disagreement over some of the ends that the UNFCCC should bring about. In Chap. 3, I argued that when there is reasonable disagreement over such ends, and when there is a pressing need to reach agreement amongst a collective group, procedural values become additionally important in the UNFCCC. In Chaps. 4, 5, 6, and 7, I then introduced several principles of procedural fairness and showed what measures are needed in order to translate these principles in the decision-making processes of the UNFCCC. During these chapters, I also drew attention to the fact that procedural fairness sometimes conflicts with other goals by introducing procedural obstacles that can prohibit agreement. Most importantly I drew attention to the fact that, whilst procedural fairness requires the representation of all affected actors, incorporating a sufficient diversity of actors into decisions to meet this requirement can lead to procedural difficulties. In addition to this, the arguments that I've presented so far have been based on an assumption that the UNFCCC is a comprehensive agreement that aims for the universal membership of states on a global scale.

In light of the possible procedural trade-offs associated with fair procedures, it's worth relaxing this assumption in order to consider whether fair procedures are still important for climate change institutions that have limited, rather than comprehensive membership. That is, in this chapter, I consider the role that multilateral agreements outside of the UNFCCC can play in addressing climate change. The current debate over the proper function and future of the UNFCCC makes this undertaking extremely relevant. Several authors have criticised the UNFCCC's comprehensive approach, arguing that there are too many divergent interests among its participants for it to bring about meaningful action on climate change. These authors argue that it might be better to focus efforts to address climate change in other areas instead. A cooperative arrangement amongst a small group of

L. Tomlinson, *Procedural Justice in the United Nations Framework Convention on Climate Change*, DOI 10.1007/978-3-319-17184-5_8

likeminded actors might be a much more conducive forum for achieving meaningful action on climate mitigation. Such proposals appear increasingly desirable in light of the pressing need to address climate change and the lack of action in the UNFCCC so far.

Contrary to these claims, I ultimately argue that international efforts to address climate change should remain under the remit of the UNFCCC. This is because addressing climate change requires sustained action on a global scale. This, in turn, requires global support from a very wide range of actors, which is dependent on a procedurally fair agreement. I argue that, by virtue of its comprehensive membership and inclusiveness, the UNFCCC is the only forum that can provide adequately fair representation and participation in its decision-making processes. Therefore, the UNFCCC is the only suitable forum for addressing climate change at the global level. As such, contrary to many existing arguments on this matter, I argue that international efforts to address climate change should continue to operate primarily through the UNFCCC. At the same time, I argue that different multilateral approaches to climate change are not mutually exclusive, and that minilateral agreements can contribute to solving the problem provided they operate in line with the principles and outcomes of the UNFCCC.

In summary, I make five central claims in this chapter.

(1) Avoiding dangerous climate change is extremely important
(2) In addition to stringency and urgency, avoiding dangerous climate change requires action that is: (i) sustained and (ii) comprehensive
(3) There is a potential trade-off between designing climate institutions that are procedurally fair, and those that achieve collective action quickly
    However,
(4) Long-term and sustained cooperation on climate change on a comprehensive scale depends on procedural fairness
    Therefore
(5) Multilateral efforts to address climate change should primarily operate through the UNFCCC

8.2 Procedural Trade-Offs

In Chap. 3, I argued that the ultimate goal of the UNFCCC is to stabilise atmospheric concentrations of greenhouse gases at a level that avoids causing dangerous climate change. Further, I suggested that there is a general understanding that meeting this goal requires limiting the increase in global temperature to no more than 2 °C above preindustrial levels. This threshold is important because a global temperature increase beyond this level is associated with extremely bad outcomes. It is thought that exceeding this level may produce potential feedback effects or exceed tipping points that bring about irreversible changes to the climate. Given the broad support for the 2 °C target, as well as the negative effects associated with a temperature increase beyond this level, it seems reasonable to suggest that this target should be used as a overall gauge for the effectiveness and desirability of global efforts to address climate change.

The problem is that meeting the demands of procedural fairness can be extremely costly for other important ends, including the ability of an institution to avoid dangerous climate change. That is, procedural fairness sometimes puts constraints on a group's ability to make meaningful decisions, and to do so quickly. Given the importance of urgency, this is very important for avoiding dangerous climate change. If the members of a multilateral group are unable to take meaningful action quickly then they will fail to adequately address climate change. Given that this book has proposed several measures for achieving procedural fairness in the UNFCCC, it's worth considering whether these measures should be overlooked for the sake of urgency. For this reason, the rest of this section sets out some of the potential trade-offs that arise with fair procedures. The subsequent section then considers whether these trade-offs warrant a multilateral approach that forgoes procedural fairness for the sake of more immediate action.

It's easiest to see how procedural fairness can conflict with other ends by considering some of the requirements for fair participation that I outlined in Chap. 4. This chapter suggested that, according to fairness, the UNFCCC should incorporate a large number of participants in its decision-making processes, so that all those who are affected by climate change are represented in its decisions. At the same time, incorporating enough actors into a decision-making process to meet this requirement is difficult on a practical level. Bringing more voices into a debate means that decisions can take longer, are more complicated, and are more susceptible to obstruction from uncooperative actors. This means that, generally speaking, including more actors in a decision prohibits a group's ability to reach agreement and make decisions quickly.

But beyond its implications for urgency, there are also other reasons for thinking that procedural fairness may impede action on climate change. For example, in Chap. 4, I also argued that procedural fairness requires taking into account the views and judgements of a diversity of stakeholders in a decision. But accommodating the interests of all relevant actors can bring about an outcome that is less stringent than one that does not. It might be that some actors are incapable of realising what their best interests are, or that their decisions are dominated by short-term views. If people are poor decision-makers, or if there is wide disagreement about the correct course of action needed to avoid dangerous climate change, then accommodating all of these views and judgements in a decision may be detrimental to reaching the ultimate outcome that we want. This implies that accommodating procedural fairness may bring about less concerted action on climate change. This is a much heard argument in the literature on environmental democracy, where there are frequent claims that democratic decisions are not necessarily those that promote the most desirable environmental ends.[1]

[1]Barry Holden, for example, discusses this point: Holden 2002.

If procedural fairness wasn't important, then an expert group could make these decisions, or decision-makers could exclude those whose opinions do not bring about the most desirable outcome. But doing this goes against the requirements for procedural fairness set out in the earlier parts of this book, which imply that actors should endeavour to treat other reasonable actors in a reasonable way, seeking to find mutually acceptable fair terms of cooperation. Provided that each actor meets the conditions of reasonableness that I set out in Chap. 2, actors should treat each other's views with respect, even when they think that a particular view is wrong. But the point is that accommodating each actor's views and opinions may prevent the overall group from taking more stringent action on climate change. One might argue that if a group knows that it should take a certain course of action (for example, preventing dangerous climate change), then it should push ahead with this goal rather than let itself get bogged down in an attempt to accommodate the interests of many different actors. This presents a second potential trade-off between procedural fairness and the overall goal of avoiding dangerous climate change.

A third trade-off occurs when decision-makers are poorly representative of those that they act on behalf of. Earlier in this book, I argued that certain actors have the right to participate in the decisions of the UNFCCC by virtue of their democratic mandate. That is to say, procedural fairness means that decision-makers should participate in decisions on behalf those that they are appointed to represent. These actors gain their legitimacy at least partly because they represent some constituency. Procedural fairness therefore also requires that these actors accurately represent those that they claim to act on behalf of. As I argued in Chap. 7, this means that if a state delegation is wholly unrepresentative of its domestic population in the UNFCCC then it might be worth excluding that actor from the UNFCCC's decisions. After all, a state gains its legitimacy as a decision-maker in the UNFCCC in part because it acts on behalf of those who have a right to participate in the decision and if it isn't acting on behalf of these actors, then it is worth questioning its place in the decision-making process.

But a problem arises if a large emitting state is wholly unrepresentative of those that it has a right to participate on behalf of. State accountability can be very weak at the multilateral level and it's not too difficult to imagine situations where a member state delegation participates in the decisions of the UNFCCC in a way that is contrary to the best interests of those that it is supposed to act on behalf of. In this case, procedural fairness may require the exclusion of that state from the decision-making process. But successfully avoiding dangerous climate change depends on reducing the emissions of all major emitting states. Assuming that any state that is excluded from an institution's decisions is unlikely to comply with any subsequent institutional commitments, this means that there is a third trade-off between the requirements of procedural fairness and successfully avoiding dangerous climate change. This arises because procedural fairness requires the exclusion of some states even when including these states is necessary for avoiding dangerous climate change.

The account of the potential trade-offs that could arise in the UNFCCC between fairness and effectiveness is not exhaustive and there may be many other cases where creating an institution which is procedurally fair agreement is detrimental to the achievement of other ends. But the account that I've given here clearly demonstrates that there are likely to be instances where procedural fairness conflicts with other ends. Given the severity of climate change, any account of procedural fairness in the UNFCCC needs to take this problem into account. That is, it is necessary to consider whether procedural fairness can be given up for the sake of more pressing ends.

In Chap. 3, I argued that this was not a feasible option in the UNFCCC on account of the fact that procedural fairness is a necessary criterion for avoiding dangerous climate change when there is reasonable disagreement about how this should be done. This argument is based on an assumption that the UNFCCC has comprehensive membership, and that it is necessary to reach an agreement that is sufficiently acceptable to all of the states of the world. In light of the trade-offs that can arise through trying to incorporate all of these actors in a multilateral agreement, it is time to relax this assumption. That is, it is worthwhile considering whether the ineffectiveness of the UNFCCC so far warrants taking an approach amongst a limited number of actors. The next section describes what such an approach would look like and explains how it might achieve more effective action on climate change.

8.3 Bypassing the UNFCCC to Avoid Deadlock

Traditional arguments for avoiding dangerous climate change maintain that climate change is a global problem that requires a global solution. Emissions are created by actions on a global scale and any reductions in emissions in one part of the world can be undone by emissions elsewhere. This means that climate change mitigation is a global public good. In fact, if several states decide to take action on climate change without the cooperation of others, then those who are outside of the agreement may have an incentive to increase their emissions, freeriding on the mitigation efforts of others (Barrett and Stavins 2003, p. 358). This means that avoiding dangerous climate change requires the participation of a very large number of actors, if not a wholly global approach.

But the lack of progress associated with the comprehensive approach of the UNFCCC has led many to question this assumption. Whilst climate change is caused by the actions of agents on a global scale, a large proportion of global emissions is caused by only a small handful of states. It's therefore theoretically possible to achieve significant emissions reductions amongst only a small subgroup of actors. One might argue that all that's needed to bring about large reductions in global emissions is the cooperation of those states that are emitting a lot now.[2] An exclusive agreement amongst a small number of likeminded states would not face the same procedural problems suffered in the UNFCCC.

[2]I take this suggestion from Bodansky and Rajamani (2013).

In light of the many procedural problems that beset the UNFCCC, several authors have started to consider whether multilateral arrangements with only partial membership might bring about more meaningful action on climate change. This was something first proposed by Moses Naim, and subsequently discussed by Robyn Eckersley and Weischer et al. (Naím 2009; Eckersley 2012; Weischer et al. 2012). These agreements represent 'minilateral' arrangements or 'clubs' of small groups of likeminded states that seek to implement cooperative action on climate change. These represent pragmatic responses to climate change, which are more politically feasible than the overambitious approach of the UNFCCC and may be better suited to achieving action on climate change (Eckersley 2012). By limiting participation to only a small number of key actors, these exclusive institutions may avoid some of the procedural problems of reaching agreement in a large group, even if they exclude those who have a right to participate as a matter of fairness. As a result, some have suggested that this sort of pragmatic approach is more likely achieve action on climate change, rather than comprehensive, binding agreements (Victor 2001, 2010; Prins and Rayner 2007; Levi and Michonski 2010).

It's worth noting that some authors also consider similar agreements that involve both state and non-state actors.[3] Others note that there are also many bilateral laws and arrangements that coordinate climate policies between two countries (Levi and Michonski 2010). There are also many instances where action on climate change is sought at the sub-state level (The World Mayors and Local Governments Climate Protection Agreement 2012). Further, many authors also argue that action on climate change has become 'fragmented', incorporating a large number of different actors and operating through many diverse processes (Biermann et al. 2010, p. 287; Biermann et al. 2012).

Whilst this literature is important, this chapter is primarily concerned with state-based multilateralism that attempts to coordinate sufficient action to address global climate change. This is partly because it seems likely that implementing sufficient efforts to successfully avoiding dangerous climate change will depend on a level of coordination which is only possible through a state based approach at the global level.[4] Smaller institutional arraignments may play an important role in contributing to this end, but are insufficient by themselves. Further, Given that multilateral negotiations have taken place over the past two decades, any proposal for institutional reform must take the existing institutional framework into account, because many political constraints arise due to the historical contingency associated with the previous negotiations. This is not to say that broader arrangements outside state-based multilateralism are unimportant, nor is it to prioritise multilateralism in any way. In this chapter, I simply assume that state-based multilateralism is very likely to play a major role in addressing climate change in the immediate future.

---

[3] See for example: Gupta et al. 2007, p. 761; Bäckstrand 2008; Bulkeley and Newell 2010; Abbott 2013.

[4] Keohane and Grant suggest that state-based arrangements therefore provide unique benefits in comparison to other modes of governance (Keohane and Grant 2005).

8.4 'Minilateral' Agreements for Climate Change

Climate minilateralism isn't just a matter of academic debate; it's also something that's seen in practice. In light of the lack of progress in the UNFCCC, many alternative multilateral arrangements now exist to coordinate cooperative action on climate change.[5] Perhaps the most notable of these agreements is the Major Economies Forum on Climate Change and Energy (MEF), which facilitates dialogue on climate change among 17 of the world's largest emitters.[6] Other initiatives operate on regional scales, such as the Asia-Pacific Partnership on Clean Development and Climate (APP), Asia-Pacific Economic Cooperation (APEC) Forum, the Global Climate Change Alliance (GCCA), and the US Major Emitters Initiative.[7] Many of these agreements cover a multitude of policy areas, whereas others have more limited mandates. For example, the Climate and Clean Air Coalition is a global movement to reduce emissions of specific greenhouse gases known as 'short lived climate pollutants' and doesn't address other issues of climate change.[8] Action on climate change is also being pursued in small multilateral groups that traditionally have much broader mandates, such as: the Group of Eight Industrialised Countries (G8) and the Group of Twenty (G20). Given that the G8 and G20 members contribute to a large proportion of global emissions, many decisions within these institutions have direct, and indirect implications for climate change.

It's clear that, whilst there has been little meaningful action under the UNFCCC, action on climate change is being pursued in other multilateral arrangements. From the perspective of climate change, the primary appeal of these institutions and agreements is that they allow likeminded actors to take action on specific issues without having to reach agreement in the UNFCCC. To be sure, there's no certainty that these approaches will implement sufficient action to avoid the very worst consequences of climate change. There may still be reasonable disagreement within these smaller groups of actors. Several studies show that unilateral measures are inadequate to prevent dangerous climate change (IEA 2011; Rogelj *et al.* 2011; UNEP 2013), and there is no promise that there will be effective agreement among a small subgroup of nations.[9] Moreover, even if there is agreement amongst a select group of key states, there is no guarantee that minilateralism can bring about the necessary changes to address climate change.[10]

[5]Stavins et al. 2014; For more on climate change initiatives outside of the UNFCCC, see: Pattberg and Stripple 2008; Biermann 2010; Keohane and Victor 2013; Bulkeley and Newell 2010.

[6]See: The Major Economies Forum on Energy and Climate 2013.

[7]For more on regional partnerships, see: Höhne et al. 2008; Kulovesi and Gutiérrez 2009. For more on the Asia-Pacific Partnership, see: APP 2012.

[8]See: UNEP 2012.

[9]Robyn Eckersley argues that much of the disagreement that exists between states on a comprehensive scale also exists between the major emitting states (Eckersley 2012, p. 33).

[10]For accounts of why it is necessary to incorporate all states, see: Hahn 2009, p. 569; van Vliet et al. 2012.

At the same time, comprehensive approach of the UNFCCC fares little better. It's well documented that the voluntary pledges that states have made in the UNFCCC so far are inadequate for avoiding dangerous climate change (UNEP 2013). The possibility that more exclusive multilateral agreements can avoid some of the problems of the UNFCCC makes it worthwhile to explore these approaches.

It's worth noting that not all those who advocate these minilateral agreements are in favour of giving up on the UNFCCC. These institutions are not mutually exclusive, and addressing climate change may require a combination of initiatives. But there are at least some who suggest that the UNFCCC is a redundant institution that uses up political resources that could be better used elsewhere (Naím 2009; Prins et al. 2010). Given these arguments it is worth considering whether the discussion so far can throw any light on the merit of continuing to support action in the UNFCCC.

Contrary to these claims, I propose that the UNFCCC is the only suitable forum for addressing climate change. I do this in three steps. First, I argue that any efforts to adequately address climate change will require action on a long term and sustained basis. Second, I argue that there are good reasons for supporting a comprehensive approach to multilateral action on climate change on the basis of efficiency. This, in turn, requires public support for an institutional arrangement *on a global scale*. Third, I argue that, on account of the fact that it is the only forum on climate change that seeks the representation of actors on a global scale, the UNFCCC is the only appropriate venue for achieving these goals. This is not to say that the UNFCCC is currently globally representative in this respect; many reforms may be necessary to meet the standards set out in Chaps. 4, 5, 6, and 7. But the UNFCCC is currently the only forum that has universal membership of all actors at the global level and seeks to accommodate principles of fairness in its decisions. In light of this, I now explain why a sustained and comprehensive approach to climate change is needed.

8.5 Sustained Cooperation

In addition to requiring urgency, avoiding dangerous climate change also requires long-term, sustained action. It is therefore important to think about how cooperation can be sustained in the long run whilst also achieving immediate action now. There are at least two points that support the view that sustained action is important.[11]

Emissions are important whenever they are created. Assuming that people want to prevent dangerous climate change from ever occurring, it does not matter whether a unit of emissions is generated today, in a year's time, or in 50 years time. In terms of the overall effect on the climate, what matters is the total cumulative amount of emissions that are generated, rather than the rate at which they are emitted, or the point in time that they are produced. One might argue that postponing emissions

[11]For a thorough account of the need for sustained action see: Dirix et al. 2013, p. 5.

now and emitting them at some point in the future avoids causing damage to the climate in the interim. Whilst this is true, the point is that it is important to avoid *ever* exceeding the 2 °C threshold, rather than seeking some temporal respite from climate damages. This means that any mitigative action in the short-term can be quickly undone by action in the future. Whilst mitigative action today may prevent the harmful effects of climate change from happening in the short-term, unless people refrain from creating emissions on a sustained basis then climate change will simply be delayed until a later point in time. Consequently there is little point in taking action in the immediate future without ensuring that this action takes place in the future as well.

A further reason for sustained action is that it may take a very long time to implement the necessary mitigation measures needed to address climate change. Carbon emissions have become a fundamental part of almost every aspect of society. It is generally thought that reducing global emissions enough to avoid the very worst effects of climate change requires large-scale changes to the global economy and to global energy infrastructure. The problem is that this is unlikely to be achievable in a short time frame. Implementing the necessary measures to address climate change requires the development of new technologies and infrastructure over a very long time period. This is only possible through sustained efforts to bring about large-scale changes to society. Therefore sustained action is needed to bring about the necessary societal changes to avoid dangerous climate change.

It might also take a long time to implement mitigation measures because of the amount of time needed to negotiate and design a successful mitigation agreement. Designing a successful climate agreement among states is an extremely complicated and challenging task. Bodansky and Diringer argue that, rather than taking the necessary time to develop a cooperative arrangement climate change, the global community has so far tried to implement a cooperative framework for the climate as quickly as possible (Bodansky and Diringer 2007). This has meant that key issues have been left unresolved and have caused problems at later points in time. Barrett and Stavins also criticise the negotiation process of Kyoto Protocol on these grounds, arguing that it sought immediate benefits without taking into account how participation and compliance could be achieved in the long-term (Barrett and Stavins 2003, p. 353). Successful climate mitigation therefore requires a long-term approach on account of the time needed to design a successful agreement. Given these two reasons, it is necessary to implement regulation across a sustained period of time, rather than over a short-term period.

8.6 Comprehensiveness

In Chap. 3, I argued that efforts to avoid dangerous climate change require limiting an increase in global temperature to no more than 2 °C above preindustrial levels. I also argued that this requires action that is urgent. In this section, I add to this by arguing that avoiding dangerous climate change also requires action on a near global scale *as a matter of effectiveness*. That is, in order to stay within the 2 °C limit, action is needed on a very broad, if not global scale. I call this the *effectiveness argument* for comprehensiveness. This might not entail a fully global approach that incorporates every single actor, but it does require at least a very large proportion of the world's actors. This, in itself, does not support a fully global approach to addressing climate change. Rather, I provide a full argument for a global approach once I've shown that climate change requires global public support, which I undertake in the following section. But turning back, for the moment, to the effectiveness argument for comprehensiveness, there are at least four arguments in support of this view.

The first argument for comprehensiveness follows from the requirement for sustained action that I outlined above. That is, long-term action on climate change requires a comprehensive approach because the state of global geopolitics is not static. This means that many states that are not large emitters now may become large emitters in the future. Any efforts to reduce emissions amongst a limited number of actors today can be undone by the efforts of others in the future. In this respect, taking a short-term view that only considers the current geopolitical situation is likely to be ineffective in the long-term and a successful climate agreement needs to take a long-term approach to climate policy. As a result, implementing action amongst a small group of large emitters will be ineffective if those countries that are low emitters increase emissions in the future.

In fact, it seems likely that many developing countries that have low emissions now will become major emitters in the near future (Hahn 2009, p. 569). This is because, whilst many developing countries have low emissions now, they are expected to undergo rapid economic growth in the next few years. According to the IEA, the bulk of the increase in global energy demand over the next two decades is expected to come from non-OECD countries (IEA 2011). Given the lack of clean energy options for meeting this demand, most of this energy will be supplied from fossil fuel sources. The implication of this is that it is impossible to make significant cuts to global emissions in the long run without the involvement of many of the world's developing countries.[12] Whilst it's sometimes argued that a successful climate change agreement need only take into account the major emitters of the world because these actors produce the bulk of global emissions,[13] this argument overlooks the fact that many states with low emissions now have the potential to become major contributors to climate change in the future.

Strictly speaking, this point necessitates a dynamic approach to climate change as much as it does a long-term and comprehensive perspective. For example, one might propose that an effective agreement should incorporate states as they become large emitters, rather than starting by including every state that could potentially become a large emitter. But the problem with this approach is that it ignores the long-term damaging effects of initially excluding actors from a cooperative agreement.

---

[12]For support of this, see: Keohane and Victor 2013, p. 106.

[13]For example: Prins and Rayner 2007; Prins et al. 2010.

Unfairly excluding actors from decisions that they have a right to participate in will be severely detrimental for the overall trust and legitimacy of an agreement. The concern is that if actors are disgruntled about an unfair agreement at the outset, then it might be very difficult to gain their support for the agreement further down the line. It seems reasonable to suggest that any state that is unfairly excluded from an initial multilateral agreement is less likely to see the merits of participating in that agreement in the future. After all, actors are more willing to support institutions and agreements that they perceive as fair. Further, an institution that initially excludes actors may eventually take measures to rectify this lack of fairness. But an agreement that is initially crafted in an unfair way leaves little to be optimistic about for those who hold concerns over fairness. This suggests that measures for collective action on climate may prove ineffective if the global context in which they are developed change in the future.

A second argument for supporting comprehensiveness is that unilateral mitigation policies are unlikely to be effective if they do not involve a large number of actors due to the problem of carbon leakage.[14] This is because domestic mitigation policies have to achieve emissions reductions in a world in which there is a large international trade market of carbon-embodied goods. Carbon-embodied goods are goods whose production involves the emission of greenhouse gases. Often, a carbon-embodied good is produced in one state and then consumed in another. This is the case, for example, if the UK imports cars that are produced in a factory in Japan, or if the US imports televisions from China. But a problem arises if these goods are traded between states that have different mitigation policies. If one state enacts a carbon mitigation policy, such as a carbon tax, then this policy is undermined if carbon-embodied goods are imported from another state where they are not subject to the same tax. In fact, in a world of differential mitigation policies, implementing a carbon policy in one state might force its manufacturing sectors to shift to states that do not enforce the same policy. This is known as carbon leakage, and it is a serious issue for enacting successful mitigation policies when there are differential regulations across the global market.[15]

Some authors have proposed ways around this problem, including the use of border carbon adjustments, which put tariffs on imported carbon embodied goods (Kuik and Hofkes 2010; Monjon and Quirion 2011). But, as of yet, there are no proven measures for addressing this issue. Although this is just one problem of trying to reduce carbon emissions, it does show that there are issues with implementing effective mitigation measures when there is only partial participation in an institutional framework.

These two arguments give strong support for a comprehensive, as well as a sustained approach to climate change. But there are at least two further points that can be made in support of a comprehensive approach. On one hand, it is

[14]For more on carbon leakage, see: Droege 2009; Eckersley 2010.

[15]For discussions of carbon leakage, see: Steininger et al. 2014; Roser and Tomlinson 2014. For an account of the amount of carbon embodied in global trade, see: Peters and Hertwich 2008.

well recognised that many of the world's cheapest mitigation options exist in the developing world (Hahn 2009, p. 569). This is because much of the energy infrastructure in the poorer parts of the world is inefficient, or based on highly polluting energy sources, such as coal. As a result, there are significant mitigation opportunities that are much cheaper than pursuing mitigation in the industrialised world alone.

But a further issue is that the developing world holds a vast proportion of the world's forests (Streck 2008). This is a problem because deforestation and land degradation contribute to a large proportion of total global emissions and are likely to contribute even more so in the future. Forests play an important role in the carbon cycle and deforestation is a significant contributor to climate change. The expected increase in population growth and associated increase in the global food demand are expected put significant pressure on the world's forests as people convert forested areas into farmland for food production. Given that many of the world's forests are in developing countries, it is necessary to take a comprehensive approach to climate change in order to ensure that this is taken into account in an overall institutional framework.

These points are good reasons for thinking that avoiding dangerous climate change requires a comprehensive, or at the least, very broad institutional approach. But this doesn't mean that a climate change agreement should be fully comprehensive to effectively take action to avoid dangerous climate change. Whilst there are some states that are low emitters now, yet have the potential to become high emitters in the future, there are some low emitting states for which this is clearly not the case. Some states have extremely small populations. The populations of states such as Tuvalu and Nauru number only several thousand. Even some states with sizeable populations seem very unlikely to become significant emitters anytime soon. Some states suffer form a severe lack of institutions and effective means of governance. Even the most optimistic forecasts of economic growth wouldn't lead to predictions that these states will become significant emitters any time soon. Further, whilst it is true that some of these countries contain large forests and other carbon sinks, it certainly isn't the case for all of these actors. Whilst the arguments that I've given here provide a justification for a very broad approach to climate change, they do not provide a reason for a fully global approach. That is to say, on the basis of the effectiveness argument for comprehensiveness, there is no reason for not excluding those states who are neither large emitters now, nor very likely to be so in the future from a climate change agreement.

This is an important point concerning the ideal form that a climate change agreement should take in order to avoid dangerous climate change. I've suggested that avoiding dangerous climate change requires urgent action on a long-term basis, and on a very large scale. But if incorporating large numbers of actors is procedurally problematic then it might be worth excluding those who are not needed (and never will be needed) to bring about successful action on mitigating climate change.

In what follows, I argue against this view on the basis that excluding these actors is unfair. This is problematic because successfully avoiding dangerous climate change requires the support and endorsement of any institutional response by a very large number of different actors globally. I argue that procedural fairness is a fundamental part of achieving this support. As such, procedural fairness is unavoidable for the design of effective climate institutions, and this value cannot be given up for the sake of achieving more immediate action now.

8.7 Global Public Support

Climate change is a problem caused by the decisions and actions of actors on a global scale. That is, whilst this book adopts a statist focus and analyses the behaviour of states in multilateral agreements; climate change is caused by the actions and decisions of a much wider group of actors. The decisions of individuals, communities, corporations and many other actors all contribute to climate change. The implication of this is that, if avoiding dangerous climate change requires large reductions in global emissions, then achieving this goal requires action on the part of a very large diversity of actors.

Further to this point, this action is likely to entail significant costs for these actors. The consumption of fossil fuels is an embedded and fundamental feature of the global economy and carbon emissions are created by actions in almost every aspect of society. As I argued in Chap. 3, stabilising atmospheric concentrations of greenhouse gases at a level that avoids dangerous climate change will require very stringent reductions in global emissions. Undertaking sufficient reductions of emissions to do this will entail very large costs. The upshot of this is that taking action to avoid dangerous climate change will require many different actors to undertake significant costs. This is not just a matter for states; it is also an issue that affects very many actors in society.

It is true that the decisions that states make about climate change at the multilateral level have serious implications for actors both domestically and internationally. States have the power to set laws and regulations for climate change that domestic actors have to comply with. The decisions that states make in a domestic context will also have many implications for actors globally. But the point is that effectively avoiding dangerous climate change requires a large diversity of actors to reduce emissions and bare any associated costs. This is important because, whilst states can impose some forms of regulation on actors both within and outside their borders, getting actors to undertake these actions requires their endorsement and support. In what follows, I outline three reasons why this is the case.

For one thing, most states are at least partly accountable to their domestic constituencies. States may participate in multilateral agreements and set domestic laws, yet these actors are often ultimately accountable to their populace. Any regulations that a particular state government puts in place may be rolled back if its electorate chooses to elect a government that reforms those measures. This is something that's become particularly evident in Australia's government over the past 2 years, where initially strong environmental measures have been repealed under a government with fewer interests in carbon mitigation. The point is that, unless there is broad support for a multilateral institution from those to whom a state is accountable, then that institution's actions may ultimately prove redundant.

A second point is that, whilst states can impose carbon regulations on domestic actors, compliance and enforcement mechanisms for these regulations are imperfect at the domestic level. Some private actors may be able to operate outside the jurisdiction of states, by easily switching to a different policy area if a regulation is put in place. The most obvious example of this is the recent international tax avoidance problems caused by corporations operating in multiple countries. This problem could also arise if individuals have the ability to emigrate to other countries with less stringent carbon policies. Even in a domestic context, one might conclude that the proper regulation of carbon emissions is a very difficult task. Given the many different day-to-day activities that create emissions, adequately accounting for emissions on an individual scale will be extremely difficult, and potentially subject to abuse. If compliance and enforcement is not comprehensively effective at the national level, then states may not be able to force actors to undertake commitments and comply with regulations all of the time.

These two reasons represent necessary reasons why public support for an international institution is important. Now, I consider a third point which suggests that broad support is an important, but not necessary, element of a climate change agreement. Above, I suggested that most states are at least partially accountable to their domestic constituencies. At the same time, there are also some states for which this isn't the case. Further, in either of these eventualities, states may, for whatever reason, fail to take sufficient action on climate change. State actors may pursue more short-term interests rather than adopt costly climate change commitments. They may hold ideological views against this sort of action. This is true for state actors that have some accountability to their citizens, and for those that do not. But in these cases, non-state actors can play an extremely important role in delivering action on climate change, even in the absence of state action. The absence of action on climate change at the state level doesn't preclude collective action through domestic actors. In fact, there are now many well-documented instances of domestic actors taking action on climate change despite the lack of action at the state level.[16] This suggests that some action on climate change can be taken even when some states are reluctant to engage in a cooperative agreement to reduce emissions. This action is most likely to arise where there is strong support amongst that state's domestic actors for a global mitigative effort. It seems unlikely that these actors would undertake action to support such an institutions efforts if this were not the case. This provides a further reason for thinking that broad support for an institution is an important part of achieving action on climate change.

[16]For discussion, see: Bulkeley and Newell 2010.

8.8 The Necessity of Procedural Fairness for Climate Change

These points provide reasons for thinking that broad public support for a multilateral agreement is necessary for addressing climate change. Further, following my earlier arguments, this public support needs to be long-term and widespread. Here, I suggest that the only way that an institution is likely to gain this sort of endorsement from a sufficient numbers of actors to implement effective action on climate change is if the institution is perceived as fair. This is because actors are generally only willing to provide long-term and unconditional support to those institutions that they think are fair. Whilst it might be possible to gain an actor's cooperation in an unfair institution in the short-run by, for example, providing sufficient incentives for participation; that actor is unlikely to continue supporting the institution if the circumstances that create those incentives change. Given that procedural fairness is an important element of fairness, and given that actors are likely to reject the commitments of an institution that does not respect this value in the long-run, procedural fairness is a critical part of a successful multilateral agreement on climate change, where success is judged by the ability to bring about sufficient action for avoiding dangerous climate change. The fact that climate change is an issue characterised by uncertainty and a need for long-term action reinforces this point. In situations where full compliance is very difficult, fairness can provide one way of gaining the necessary support for the agreement.

This suggests that fairness is a necessary element of any multilateral agreement that seeks to mitigate emissions at a level that avoids dangerous climate change. Procedural fairness is an important part of the overall fairness of an institution and, as such, addressing dangerous climate change requires a procedurally fair institution. But further to this, my discussion from Chap. 2 gives additional importance to procedural fairness here. If, as I have shown, there is reasonable disagreement over the ends that a climate institution should pursue, then procedural fairness can provide a way of reaching an agreement that all find sufficiently fair to be acceptable.

The upshot of this is that effectively avoiding dangerous climate change also requires an institutional approach that is procedurally fair. This means that it's now possible to think about what this means for the design of a multilateral agreement for climate change. But before doing so, it is worth briefly recalling the argument so far.

Earlier in this chapter, I made the following claims:

(1) Avoiding dangerous climate change is an important end
(2) Achieving this end means implementing action that is: stringent, urgent, sustained, and comprehensive
(3) There is sometimes a trade-off between procedural fairness and achieving action quickly

In this section, I then made the following claim:

(4) Long-term and sustained cooperation on climate change on a comprehensive scale depends on procedural fairness

Having shown this, it is now time to turn to the fifth claim of this chapter:

(5) Multilateral efforts to address climate change should primarily operate through the UNFCCC

8.9 The Primacy of the UNFCCC for Addressing Climate Change

It's now time to think about what this all means for the UNFCCC, and for attempts to coordinate an international response to climate change more generally. Earlier, I noted that 'minilateral' or exclusive multilateral agreements have been proposed as a way of avoiding some of the deadlock and inaction in the UNFCCC. I also noted that many of these arrangements now exist at the international level, and that it is evident that states are form 'clubs' that are sufficiently large to take serious steps towards reducing emissions. Further to this, I noted that some authors have begun to question whether the international community should continue supporting the UNFCCC and its ultimate aim to create a comprehensive coordinated response to climate change. Each major UNFCCC conference is preceded by unrealistic media hype and political attention that creates idealistic expectations about what the international community can achieve. The subsequent disappointment that follows the failure to bring about action puts a heavy toll on public support and hope for successful action on climate change. The UNFCCC also carries a large political cost, eating away at limited resources that states could potentially put into other, more efficient agreements. Given these costs and given the possibility of promoting action elsewhere, it's certainly worth considering whether it might be time to give up on the UNFCCC altogether.

Yet in spite of these costs and doubts, I propose that international efforts to address climate change should continue to support action within the UNFCCC. Contrary to the arguments for giving up the UNFCCC, the discussion so far suggests that if procedural fairness is a critical part of a climate change regime that avoids dangerous climate change, then the UNFCCC is also critical to meeting this goal. This is because the UNFCCC is currently the only agreement capable of making decisions that meet the demands of procedural fairness that I outlined earlier in this book. That is to say, the UNFCCC is the only institution that explicitly incorporates principles of procedural fairness in its design. It attempts to incorporate all states into its decision-making processes on a comprehensive basis. It also seeks the inclusion and participation of a variety of non-state actors that act on behalf of stakeholders who are poorly represented in the state-based system. This is important because, as I argued in Chap. 4, the representation of all those affected by a decision is an important part of a fair decision-making process. Further, the UNFCCC actively seeks to address some of the procedural injustices that arise when decision-makers have unequal resources. For example, the UNFCCC actively assists those states with limited financial and technical resources for participating in decisions, thereby promoting the ideals of political equality that I defined in Chap. 5. To date, other multilateral institutions that attempt to address climate change at the global level fall short in this respect. Many of these institutions operate in closed forums, without the participation of the non-state actors or media representatives.[17] If procedural fairness is an important part of addressing climate change, then this means that the UNFCCC is the most appropriate multilateral arrangement for addressing climate change in the immediate future.

To be sure, procedural fairness and proper representation in the UNFCCC is still far from ideal. Many states are poor representatives of those that they act on behalf of. Many non-state actors can participate in decisions without any sort of accountability to those that they claim to represent. Whilst the UNFCCC takes some measures to level the playing field between different delegations, there is still a great deal of inequality between these actors. Nevertheless, it remains the fact that the UNFCCC is institution for addressing climate change at the global level that is a public body which aims to incorporate the interests of actors on a global scale in a fair way. Whilst the UNFCCC is not absolutely fair at this point in time, it remains the most likely body that can fulfil the necessary demands for procedural fairness that I've described in this book. This is what gives it greater legitimacy than its multilateral rivals. It might ultimately be the case that a minilateral arrangement takes a procedurally fair approach by incorporating a large number of voices under fair terms. But this seems far form the case at the moment, where many of these institutions adopt an exclusive approach to decision-making, deliberately limiting decisions to only a small number of actors. For all its flaws, the UNFCCC remains the most likely candidate for providing the necessary procedural fairness to bring about the public support needed for action on climate change. If procedural fairness is a key element of a global effort to address climate change, and if the UNFCCC is the most likely forum that can meet the requirements of procedural fairness, then the UNFCCC is the most appropriate institution for global efforts to address climate change.

8.10 Combining Institutional Approaches

But the issue of urgency still looms large here. Whilst it is important to take a long-term view towards climate change, and whilst this requires considering procedural fairness, it is also necessary to keep in mind the need for urgency. Whatever one thinks about the need for a long-term view, these arguments become trivial if

[17]The Cartagena Dialogue (Bowering 2011) the MEF, and the G8 (Karlsson-Vinkhuyzen and McGee 2013, p. 67) are closed to observers.

catastrophic outcomes aren't prevented in the interim and prioritising either urgency, or the long-term view is a mistaken approach to take. Climate stabilisation requires us to look at both of these issues, rather than giving up on either one for the sake of the other. So it's necessary to develop an institutional approach that: (i) deals with urgency in the short-term, (ii) whilst promoting long-term cooperation, and (iii) that does so in a fair way. Furthermore, whilst I propose that efforts to address climate change should continue within the UNFCCC, this doesn't rule out other institutional processes. In fact, these agreements could still play an important role in the overall effort to address climate change in a way that meets these three requirements. The remainder of this chapter now considers what sorts of institutional arrangements can reconcile this aim.

Several authors have suggested ways of combining (i) and (ii) in an institutional response to climate change. Proposals in this vein suggest taking some immediate action through minilateral approaches, whilst also focussing on long-term aims for the future. For example, Daniel Bodansky suggests that climate institutions should adopt an 'evolutionary framework', which would take some immediate action now with the view to adopt stronger measures in the future (Bodansky 2012). Arunabha Ghosh argues that the UNFCCC could take a similar approach to the World Trade Organisation, which imposed 'flexible' commitments during its initial implementation (Ghosh 2010). This involved an initial agreement amongst a limited group of actors that gradually expanded its membership and commitments over time. Johannes Urpelainen suggests that climate institutions should start on a small scale and progressively coordinate agreements between different policy areas; a key requirement being that states *demonstrate* that their participation in small-scale agreements contributes to the development of an ambitious agreement in the future (Urpelainen 2013). More recently, Grasso and Timmons Roberts have suggested that the most realistic avenue for action is to reach an initial agreement amongst major emitters in the MEF and to use this a foundation for more comprehensive action (Grasso and Timmons Roberts 2013). A common theme throughout each of these examples is that small groups combine select issues at first, with the ultimate aim of reaching a stronger agreement in the future.

These sorts of approaches combine issues (i) and (ii) above. But it is also necessary to combine (iii) fairness. That is, it is necessary to think about how climate institutions can achieve immediate action, whilst keeping a long-term focus, *in a fair way*. Here I focus on two ways of doing this. First, some authors, such as Bodansky, and Keohane and Victor, advocate an 'umbrella' approach. Rather than taking a fully comprehensive approach, Bodansky's proposal involves initially developing multilateral agreements in different sectors and among different actors in the way set out above. But for Bodansky, this should be done under the guise of a core institutional arrangement, or 'umbrella' that controls and directs these smaller agreements. Whilst Bodansky doesn't specifically mention fairness, this sort of approach would allow an institution to impose standards of procedural fairness on small-scale agreements, whilst building towards a comprehensive agreement. Similarly, for Keohane and Victor, the UNFCCC would continue to supply certain core elements of an overall institutional framework for climate change that could help guide and support the wider network of 'clubs' that coordinate action on climate change (Keohane and Victor 2010).

Second, some advocate the role of participation, trust and equality in bringing about institutional arrangements that are sufficiently flexible to adapt and adopt new commitments over time. A key part of doing this is maintaining the participation of non-member states and high levels of transparency. For example, Ghosh argues that the evolution of the WTO benefited from 'progressive multilateralism' and that the UNFCCC should take a similar approach (Ghosh 2010). This meant that some states contributed to its design even though they weren't subject to its full commitments. Likewise Dirix et al. argue that an important part of flexible institutions is keeping external actors involved in the development of small scale institutions, in order to develop sufficient trust to arrive at a comprehensive agreement in the long-run (Dirix et al. 2013). Robyn Eckersley attempts to reconcile the benefits of minilateral agreements with those of comprehensive climate multilateralism (Eckersley 2012). Eckersley defends a form of inclusive minilateralism, which coordinates collective action amongst a key group of major emitters, whilst a 'Climate Council' of those who are most affected by climate change participates in its design and implementation. This allows immediate action, whilst maintaining a sufficiently fair process to develop a comprehensive agreement in the long-run. Dryzek and Stevenson also attempt to bring a more inclusive approach to minilateral agreements through deliberative democracy (Dryzek and Stevenson 2012).

But the important thing to note from this is that the institutional approaches are not mutually incompatible; multilateral efforts can continue in the UNFCCC *and* in other minilateral forums. But it is wrong to suggest that the international community should give up on the UNFCCC for the sake of pursuing action elsewhere, which is something that's been proposed in the literature on this subject.[18] Rather, these different approaches should be designed so that they are mutually reinforcing, coordinating action to achieve the overall goal of stabilizing greenhouse gas concentrations. Further, whilst global efforts to address climate change can take place in a number of different fora, it is important that this is done in a fair way. Following the arguments that I've made in this chapter so far, the UNFCCC should act as the overarching institution that provides the primary forum for discussion on climate change. Whilst this shouldn't preclude the opportunity for states to cooperate in small groups before implementing action in the UNFCCC, these smaller groups shouldn't act as alternative negotiation arenas for states to bypass issues that cannot be resolved in the UNFCCC. Improving the inclusiveness of minilateral approaches may help to make these arrangements fairer in a procedural sense. But the UNFCCC can provide a primary avenue of discussion on climate change which ensures that each and every actor's interests are represented in the debate in the first place.

[18]For example: Naím 2009.

This also means that urgency and sustained cooperation aren't mutually incompatible aims and that an institutional approach should strive to achieve both of these simultaneously. At the same time, fairness also plays an important role as a necessary feature of sustaining long-term cooperation. All of these issues should be kept in mind when thinking about how to design procedures. If there are trade-offs between long-term outcomes and more urgent goals, then international efforts should attempt to resolve these issues to any extent possible, rather than giving up on either one. Of course, there are concerns that seeking procedural fairness may prove too time consuming in light of the time constraints for limiting emissions.[19] But what this chapter has shown is that giving up on procedural fairness for the sake of urgency is only likely to provide a temporary solution to the problem. I've also given some examples of how these competing ends can be met simultaneously. Even if continued delay makes it worth pursuing climate governance through institutions outside the UNFCCC, this is not to say that the actions taken by the UNFCCC are no longer crucial for addressing climate change.

Throughout this chapter, I've emphasised the intractable nature of procedural values for the overall design of climate institutions. This is not to say that procedural fairness is an absolute value, or that it should be prioritised above other values. Rather, it is simply intractable from other elements of institutional design, given that it is an important part of climate stabilisation. I've also made some preliminary suggestions about how best to design an institutional response to meet these ends. In doing so, I've made an argument for maintaining support for action on climate change through the UNFCCC on the grounds that it is a procedurally fair institution that can help guide the formation of other multilateral mechanisms for addressing climate change.

8.11 Summary of Policy Recommendations

This chapter has drawn together the separate threads of this book to provide an account of why the UNFCCC is a fundamentally important institution for any multilateral endeavour to address climate change. I've made this claim on the back of the preceding arguments that I've set out in the earlier stages of this book. It's now worth briefly concluding by summarising these main points and showing how these different arguments and policy recommendations fit together and mutually reinforce one another.

In Chap. 2, I argued that there is reasonable disagreement over the fair distribution of emission rights in the UNFCCC. I showed this by looking at most of the prominent principles of fairness that have been advocated for the distribution of emission rights both in the UNFCCC, and in discussions in academic and policy circles. By considering each of these principles in turn, I showed each could be seen

[19]For discussion: Eckersley 2012, p. 28.

as a reasonable interpretation of what's fair. The important upshot of this is that the member states of the UNFCCC are unlikely to reach agreement on this issue any time soon. Given several accompanying constraints about what's required for cooperative action in the UNFCCC (such as the voluntary agreement of all member states) I argued that this means that this means that cooperation on climate change is unlikely to come about spontaneously. I then argued that one way around this problem is to come up with a fair decision-making process that can by pass this procedural deadlock.

Having made the case for why procedural fairness is important in the UNFCCC, I then turned to the question of what procedural fairness requires in relation to climate change. I started this by asking who should participate in the decisions of the UNFCCC COP. I discussed various principles of procedural justice that we might think should determine who participates in a decision and considered the merit of each of these in relation to climate change. I discussed and rejected the most frequently referenced principle of participation, namely the All Affected Principle. I then argued that an actor should have some say in a decision if it is made in that actors name, or if the decision coerces the actor. I went on to claim that those whose interests are potentially affected by a decision have the right to express their interests and views in the way that it is made.

Having discussed *who* should participate I then turned to the question of *how* actors should participate. To answer this, I introduced some normative ideas of democracy that should guide our thinking about procedural justice. I argued that there are several principles for fair decision-making, including: the equal advancement of interests, autonomy, and justification, and I used these to develop a notion of political equality for fair decision-making in multilateral decisions. In doing so, I argued that political equality consists of two elements: equal respect and a level playing field for making decisions.

Building on this notion of political equality, I then turned to the issue of bargaining. Bargaining is one way of coming to an agreement where there are competing notions of what the fair terms of cooperation should be between two actors. Given this, in this chapter I considered what procedural justice requires for fair bargaining. I argue that fair bargains are both voluntary and reciprocal. Voluntary bargains are those that are free from manipulation and coercion, and where bargainers are informed and rational. Reciprocal bargains are those that are not exploitative, where exploitation means that one party gains a disproportionate benefit from the bargain.

An alternative approach to making decisions where there is reasonable disagreement is to vote. Voting is an important part of the UNFCCC's decisions, and the current way of voting by consensus has become a focal point for procedural criticism of the UNFCCC. This chapter discussed what procedural justice requires in relation to voting. To do this, I discussed what voting rule should be adopted according to fairness, and how votes should be weighted. I argued that voting by majority rule is one way of making decisions that are both fair and efficient. I then argued that fair voting processes are those that weight votes according to the number of people that an actor represents. Fairness also requires that votes are weighted according to the stake that these people have in a decision, provided that decisions are made for the sake of advancing each decision-maker's own interests.

This represents the principles of procedural justice that should guide the procedural design of decisions in the UNFCCC. But throughout each of these chapters, I also discussed what sort of rules and measures could be put in place. In the remainder of this section, I draw together these different strands to provide a single account of what's needed to reform the procedural rules of the UNFCCC for the sake of fairness.

A common theme throughout this book is the importance of Non-state Actors (NSAs) for procedural justice. NSAs include community groups, civil society actors, regional representatives, and Non-governmental Organisations (NGOs) such as those that represent business or environmental interests. The UNFCCC should encourage the involvement of these actors in its decisions, to the extent that they improve the overall decision-making process. These actors should have a voice in the deliberative phase of decision-making, where they can put forward views that aren't presented elsewhere. They should also be able to work alongside state delegations in order to assist them with their work and providing information about decisions to the global public. This serves two purposes. First, NSAs play a role in making sure that all views are heard in debates. NSAs fulfil a representative role, by ensuring that important viewpoints that are otherwise left out of decisions. Second, the engagement of NSAs should improve the capability of state delegates to participate in negotiations. NSAs can provide assistance to those that are unable to participate on equal terms in these processes, as well as fulfilling an epistemic role by providing additional expertise and information.

In addition to the role that NSAs can play in levelling the playing field for member states in COP negotiations, the UNFCCC should improve the communication and dissemination of its decisions to the widest audience possible. Representation in multilateral politics is often very poor. Communicating decisions to a broad audience will empower actors on a global scale, to hold multilateral institutions to account and strengthen their voice in decisions that have severe implications for people globally. Empowering people in this way will also serve to strengthen the capacity of member state delegations in COP negotiations, by providing greater political attention and motivation from the bottom up.

In relation to bargaining and voting, the UNFCCC should adopt procedural rules governing acceptable ways of making decisions. This would include allowing states to bargain, under certain constraints. For example, the UNFCCC should prohibit unjustified negative issue linkages in climate negotiations. It should also prevent decision-makers from linking issues that are detrimental to the overall negotiation process, or that allow decision-makers to exploit each other. Turning to voting, the UNFCCC should adopt procedural rules specifying that member states should make decisions by majority rule, but only after parties have deliberated all of the issues at stake. Votes should be weighted according to the number of actors that each state represents according to population size.

The UNFCCC should exclude those states that clearly obstruct negotiations. These are decision-makers that do not meet the requirements of reasonableness set out in Chap. 2. Deliberately obstructing the decision-making process fails to show an adequate respect for other decision-makers and constitutes unreasonable behaviour. Unreasonable actors make demanding claims and do not respect other actors in a decision-making process. If a decision-maker intentionally obstructs progress and agreement in climate change institutions then there are grounds for excluding that actor from formally participating in the decision-making processes of the institution.

The UNFCCC should also consider excluding states that are grossly unrepresentative of those that they claim to represent. This means that the UNFCCC should consider excluding states that have very poor democratic standards, or that clearly act against the interests of their citizens. This does not mean that states are required to meet high standards of democratic accountability in order to participate in an institution. But it acknowledges that if people care about procedural justice in climate change institutions, then they should also care about whether states act fairly towards their own citizens. In situations of blatant procedural injustice, the UNFCCC should consider excluding these actors from the decision-making body of climate change institutions.

The final message of this book is that the primacy of comprehensive, long-term cooperation on climate change is undeniable. The UNFCCC is the ideal form of cooperation in the sense that it is likely to be the most effective and the fairest. Practical constraints and feasibility issues sometimes undermine this approach. But different forms of multilateral cooperation on climate change are not mutually exclusive. Successful action on climate change will involve a combination of these approaches. The stakes are so high that our efforts and resources should be put into a number of approaches, rather than limited to any particular area of action.

References

Abbott, K. 2013. Strengthening the transnational regime for climate change. *Transnational Environmental Law* 3: 57–88.

APP. 2012. *Asia-Pacific partnership on clean development and energy*. From: http://www.asiapacificpartnership.org/english/default.aspx.

Bäckstrand, K. 2008. Accountability of networked climate governance: The rise of transnational climate partnerships. *Global Environmental Politics* 8(3): 74–102.

Barrett, S., and R.N. Stavins. 2003. Increasing participation and compliance in international climate change agreement. *International Environmental Agreements: Politics, Law and Economics* 3: 349–376.

Biermann, F. 2010. Beyond the intergovernmental regime: Recent trends in global carbon governance. *Current Opinion in Environmental Sustainability* 2: 284–288.

Biermann, F., P. Pattberg, et al. 2010. *Global climate governance beyond 2012: Architecture, agency and adaptation*. Cambridge: Cambridge University Press.

Biermann, F., K. Abbott, et al. 2012. Navigating the Anthropocene: Improving earth system governance. *Science* 16.335(6074): 1306–1307.

Bodansky, D. 2012. *The Durban platform: Issues and options for a 2015 agreement.* Centre for Climate and Energy Solutions.

Bodansky, D., and E. Diringer. 2007. *Towards an integrated multi-track framework.* Arlington: Pew Center on Global Climate Change.

Bodansky, D., and L. Rajamani. 2013. Evolution and governance architecture. In *International relations and global climate change*, ed. D. Sprinz and U. Luterbacher. Cambridge, MA/London: MIT Press.

Bowering, E. 2011. *After Kyoto: The Cartagena dialogue and the future of the international climate change regime*. Prepared for the United Nations Framework Convention on Climate Change COP17. www.globalvoices.org.au

Bulkeley, H., and P. Newell. 2010. *Governing climate change*. Abingdon/New York: Routledge.

Dirix, J., W. Peeters, et al. 2013. Strengthening bottom-up and top-down climate governance. *Climate Policy* 13(3): 363–383.

Droege, S. 2009. Tackling leakage in a world of unequal carbon prices. *Report Climate Strategies*.

Dryzek, J., and H. Stevenson. 2012. Legitimacy of multilateral climate governance: A deliberative democratic approach. *Critical Policy Studies* 6(1): 1–18.

Eckersley, R. 2010. The politics of carbon leakage and fairness of border measures. *Ethics and International Affairs* 24(4): 367–94.

Eckersley, R. 2012. Moving forward in the climate negotiations: Multilateralism or minilateralism? *Global Environmental Politics* 12(2): 24–42.

Ghosh, A. 2010. *Making climate look like trade? Questions on incentives, flexibility and credibility*, Policy brief for centre for policy research. New Delhi: Dharma Marg.

Grasso, M., and J. Timmons Roberts. 2013. A fair compromise to break the climate impasse. *Global Economy and Development at Brookings* 2013–02.

Gupta, S., D.A. Tirpak, et al. 2007. Policies, instruments and co-operative arrangements. In *Climate change 2007: Mitigation. Contribution of working group III to the fourth assessment report of the Intergovernmental Panel on Climate Change*, ed. B. Metz, O.R. Davidson, P.R. Bosch, R. Dave and L.A. Meyer. Cambridge: Cambridge University Press.

Hahn, R.W. 2009. Climate policy: Separating fact from fantasy. *Harvard Environmental Law Review* 33: 557–591.

Höhne, N., F. Yamin, et al. 2008. The history and status of the international negotiations on a future climate agreement. In *Beyond bali: Strategic issues for the post-2012 climate change regime*, ed. C. Egenhofer. Brussels: Centre for European Policy Studies.

Holden, B. 2002. *Democracy and global warming*. London: Continuum.

IEA. 2011. *World energy outlook 2011: Executive summary*. International Energy Agency. http://www.worldenergyoutlook.org/media/weowebsite/2011/executive_summary.pdf.

Karlsson-Vinkhuyzen, S.I., and J. McGee. 2013. Legitimacy in an era of fragmentation: The case of global climate governance. *Global Environmental Politics* 13(3): 56–78.

Keohane, R.O., and R.W. Grant. 2005. Accountability and abuses of power in world politics. *American Political Science Review* 99(1): 29–43.

Keohane, R.O., and D.G. Victor. 2010. *The regime complex for climate change*. Harvard Project on International Climate Agreements, Belfer Center for Science and International Affairs, Harvard Kennedy School.

Keohane, R.O., and D.G. Victor. 2013. The transnational politics of energy. *Dædalus, The Journal of the American Academy of Arts & Sciences* 141(1): 97–109.

Kuik, O., and M. Hofkes. 2010. Border adjustment for European emissions trading: Competitiveness and carbon leakage. *Energy Policy* 38(4): 1741–1748.

Kulovesi, K., and M. Gutiérrez. 2009. Climate change negotiations update: Process and prospects for a Copenhagen agreed outcome in December 2009. *Review of European Community and International Environmental Law* 18(3): 229–243.

Levi, M.A., and K. Michonski. 2010. *Harnessing international institutions to address climate change*. Council of Foreign Relations Working Paper. New York: Council of Foreign Relations.

MEF. 2013. *Major economies forum on energy and climate*. http://www.majoreconomiesforum.org/.

Monjon, S., and P. Quirion. 2011. A border adjustment for the EU ETS: Reconciling WTO rules and capacity to tackle carbon leakage. *Climate Policy* 11(5): 1212–1225.

Naím, M. 2009. Minilateralism: The magic number to get real international action. *Foreign Policy* 173: 135–136.

Pattberg, P., and J. Stripple. 2008. Beyond the public and private divide: Remapping transnational climate governance in the 21st century. *International Environmental Agreements* 8(4): 367–388.

Peters, G., and E.G. Hertwich. 2008. CO2 embodied in international trade with implications for global climate policy. *Environmental Science and Technology* 42(5): 1401–7.

Prins, G., and S. Rayner. 2007. *The wrong trousers: Radically rethinking climate policy*. Joint Discussion Paper of the James Martin Institute for Science and Civilization, University of Oxford and the MacKinder Centre for the Study of Long-Wave Events, London School of Economics.

Prins, G., I. Galiana, et al. 2010. *The Hartwell paper: A new direction for climate policy after the crash of 2009*. Oxford: Institute for Science, Innovation and Society, University of Oxford.

Rogelj, J., et al. 2011. Emission pathways consistent with a 2 °C global temperature limit. *Nature Climate Change* 1: 413–418.

Roser, D., and L. Tomlinson. 2014. Trade policies and climate change: Border carbon adjustments as a tool for a just global climate regime. *Ancilla Iuris*, November 2014. http://anci.ch/_media/beitrag/ancilla2014_roser-tomlinson.pdf

Stavins R. et al. 2014. International cooperation: Agreements and instruments. In *Climate Change 2014: Mitigation of climate change. Contribution of working group III to the fifth assessment report of the Intergovernmental Panel on Climate Change*, ed. O. Edenhofer et al. Cambridge, UK/New York: Cambridge University Press.

Steininger, K., et al. 2014. Justice and cost effectiveness of consumption-based versus production-based approaches in the case of unilateral climate policies. *Global Environmental Change* 24: 75.

Streck, C. 2008. Forests, carbon markets, and avoided deforestation: Legal implications. *Carbon & Climate Law Review* 3: 239–247.

UNEP. 2012 *All G8 countries back action on black carbon, methane and other short lived climate pollutants*. United Nations Environment Programme Environment for Development News Centre, Press release. http://www.unep.org/newscentre/Default.aspx?DocumentID=2683&ArticleID=9134&l=en.

UNEP. 2013. *Emissions gap report*. Nairobi: UNEP.

Urpelainen, J. 2013. A model of dynamic climate governance: Dream big, win small. *International Environmental Agreements* 13: 107–125.

van Vliet, J., et al. 2012. Copenhagen accord pledges imply higher costs for staying below 2 °C warming. *Climate Change* 113: 551–561.

Victor, D. 2001. *The collapse of the Kyoto protocol and the struggle to slow global warming*. Princeton/Oxford: Princeton University Press.

Victor, D. 2010. Global warming policy after Copenhagen. *Willard W. Cochrane lecture in public policy*. University of Minnesota.

Weischer, L., et al. 2012. Climate clubs: Can small groups of countries make a big difference in addressing climate change? *RECIEL*

Printed in the United States
By Bookmasters